Essays on the Land, Ecotheology,
and Traditions in Africa

Essays on the Land, Ecotheology, and Traditions in Africa

EDITED BY
Benjamin Abotchie Ntreh
Mark S. Aidoo
Daniel Nii Aboagye Aryeh

FOREWORD BY
J. Kwabena Asamoah-Gyadu

RESOURCE *Publications* · Eugene, Oregon

ESSAYS ON THE LAND, ECOTHEOLOGY, AND TRADITIONS IN AFRICA

Copyright © 2019 Wipf and Stock Publishers. All rights reserved. Except for brief quotations in critical publications or reviews, no part of this book may be reproduced in any manner without prior written permission from the publisher. Write: Permissions, Wipf and Stock Publishers, 199 W. 8th Ave., Suite 3, Eugene, OR 97401.

Resource Publications
An Imprint of Wipf and Stock Publishers
199 W. 8th Ave., Suite 3
Eugene, OR 97401

www.wipfandstock.com

PAPERBACK ISBN: 978-1-5326-8201-8
HARDCOVER ISBN: 978-1-5326-8202-5
EBOOK ISBN: 978-1-5326-8203-2

Manufactured in the U.S.A.

Contents

List of Contributors | vii

Foreword by J. Kwabena Asamoah-Gyadu | xi

Introduction

The Bible and Caring for the Land | Benjamin A. Ntreh | 3

The Bible and Environment—Care for the Land
 | Emmanuel G. L. Twum-Baah | 15

Part I

Towards an Agenda for Ecotheology in African
 Theological Studies | Mark S. Aidoo | 41

The Bible and Caring for the Land: A Theological Approach
 | Thomas Oduro | 59

The Bible and Caring for the Land: African Theocology as Christian
 Impulsion for Creation Care | Ebenezer Yaw Blasu | 70

Part II

Understanding the Idea of Ἔρημος in Luke's Gospel
 | Daniel Nii Aboagye Aryeh | 97

Salvaging Our Environment: A Reflection on the Response of the Catholic
 Church in Ghana | Bonsu Osei-Owusu | 113

Part III

Impact of Artisanal and Small-scale Mining on the Environment in Ghana | Adwoa Boaduaa Yirenkyi Fianko | 135

Exploring Possible Solutions to the Galamsey Menace in Ghana | John Appiah and James Kwaku Agyen | 146

Glocal Ecological Degradation of God's Gift: A Human Menace of the Divine Creation in Ghana | Yaw Atta Edu-Bekoe | 155

Fulani Herdsmen Traditions and Care for the Land | Haruna Yussif Mogtari | 178

Desacralization of Gold in Southwest Burkina Faso: A Christian Response to Gold Mining and its Consequences on Creation | Ini Dorcas Dah | 192

Part IV

Akan Traditional Perspectives of Land Care | Kofi Agyekum | 207

The Krobo Religious View of Land, Use, and Care | Robert Mate Wayo Opata | 219

Reflections

Reflections: A Ghanaian Christian View of Land Care | Allison Howell | 233

Contributors

Prof. Benjamin Abotchie Ntreh is a professor of Old Testament and Head of Department, Heritage Christian College, Amasaman Ghana. His specialties are Hebrew and Old Testament studies. His research interest is African Biblical Hermeneutics. His publications include *A Concise History of Ancient Israel and Judah*. He can be reached at bantreh@yahoo.co.uk.

Dr. Mark S. Aidoo (PhD) is a lecturer and Director for Graduate Programmes at Trinity Theological Seminary, Legon, Ghana. He teaches Old Testament, biblical Hebrew, and preaching. His research interests are poetic literature, African spirituality, leadership, and preaching. His recent publication is *Shame in the Individual Lament Psalms and African Spirituality*. He is the Secretary of Ghana National WAATI. He can be reached at macsaidoo@gmail.com.

Daniel Nii Aboagye Aryeh (PhD Candidate) is the Dean of the School of Theology at Perez University College, Winneba, Ghana. He is also the zonal organizer of Ghana WAATI. His specialties are New Testament studies and biblical Greek. His recent publications are *Urban Public Space Evangelism: Evangelism in Market Places in Ghana* and *Biblical, Theological, and Traditional Framework for Understanding Christian Prophetism in Ghana Today*. His research interests are New Testament studies, missions, gender studies, and pentecostalism. He can be reached at danielniiaboagyearyeh@gmail.com.

Dr. Emmanuel G. L. Twum-Baah is a senior lecturer and Vice President (students affairs) at All Nations University College, Koforidua, Ghana, where he also serves as Head of the Department of Biblical Studies, Religion, and Theology. His research interests are postcolonial biblical interpretation, biblical Hebrew, and the history of ancient Israel. His recent publication is "Interpreting the Pauline Imagery of Salvation and Adoption in Krobo Religio-Cultural Thought." He can be reached at egl.twumbaah. egltb@gmail.com.

Professor Thomas A. Oduro (PhD) is an associate professor of African Christianity. He is the President of Good News Theological Seminary, Oyibi Accra, Ghana. His research interests include the history of Christianity, theology, and African Christianity. His recent publication is *Church of the Lord, Brotherhood: History, Challenges and Growth*. He can be reached at taoduro@gmail.com.

Dr. Ebenezer Yaw Blasu (PhD) is a senior lecturer at the Centre for the Promotion of Life Values, Presbyterian University College, Abetifi, Ghana. His research interests include holistic Christian education, and African theocology. He can be reached at eyblasu@gmail.com.

Dr. Bonsu Osei-Owusu (PhD) is a lecturer in ethics and philosophy for the Department of Religious Studies and Ethics at the Methodist University College, Accra, Ghana. His research interest is environmental ethics. His publications include *Concise Notes on Business and Professional Ethics*. He can be reached at oseiowusu13@yahoo.com.

Adwoa B. Yirenkyi-Fianko is a senior lecturer at the Ghana Institute of Management and Public Administration (GIMPA), Ghana. Her recent publication is "An Analysis of Risk Management in Practice: The Case of Ghana's Construction Industry." She has a BSc in building technology and an MSc in project management. She is currently a PhD candidate (environmental engineering) at the University of Western Ontario, Canada. She can be reached at asbaah2002@gmail.com.

Dr John Appiah, (PhD) is a senior lecturer at Valley View University, Oyibi, Ghana. His research interests are Christian education and biblical theology. His recent publications include *Adventist Education and Mission in Africa* (2017), and *Integration of Faith and Learning in Distant Education in the Ghanaian Context* (2015). He can be reached at jappiah@vvu.edu.gh.

James Kwaku Agyen, (PhD Candidate) is a lecturer at the School of Basic and Bio-Medical Sciences, University of Health and Allied Sciences, Ho, Volta Region, Ghana. He can be reached at jkagyen@vvu.edu.gh.

Dr. Yaw Attah Edu-Bekoe (DMiss) is a senior lecturer at Trinity Theological Seminary, Legon, Ghana. His research interests are missiology and diaspora missiology. His recent publication is *Scattered Africans Keep Gathering: A Case Study of Diaspora Missiology on Ghanaian Migration and Congregations in the USA* (2018). He can be reached at yaw.edu-bekoe@alum.ptsem.edu.

Haruna Y. Mogtari (MA, MTh) is the Director for the Centre for Research and Islamic Studies, Akropong-Akuapem, Ghana. His research interests include Islam in Africa, and Fulani in the Ghanaian context.

Ini Dorcas Dah (PhD) is Professeur Associée (Visiteur) à l'Institut Pastoral Hébron (IPH), Bouaflé, Côte d'Ivoire and an adjunct research fellow at the Akrofi-Christaller Institute of Theology, Mission and Culture (ACI), Akropong-Akuapem, Ghana. Her recent publication is *Women Do More Work than Men: Birifor Women as Change Agents in the Mission and Expansion of the Church in West Africa [Burkina Faso, Côte d'Ivoire and Ghana]* and her research interests are Christian history, gospel and culture, and holistic mission and development.

Professor Kofi Agyekum is a professor of linguistics and the acting Dean of the School of Performing Arts, University of Ghana, Legon, Ghana. His areas of interest are pragmatics, discourse analysis, ethnography of communication, stylistics, semantics, lexicology, translation, oral literature, and terminology. He is a renowned Akan scholar and has contributed significantly to the development of the Akan language in both the academic and broadcasting fields. He can be reached at kagyekum@ug.edu.gh.

Robert Opata Mate-Wayo (MTh) is a lecturer at All Nations University College, Koforidua, Ghana. His research interests are African Christianity and African ethics.

Professor Allison Howell (PhD) was born in the Democratic Republic of Congo and is an Australian. She has worked in Ghana since 1982. She is an associate professor and adjunct lecturer at the Akrofi-Christaller Institute of Theology, Mission and Culture, Akropong-Akuapem, Ghana. She can be reached at amhowell@hotmail.com.

FOREWORD

"The Earth is the Lord's": Mainstreaming Ecological Issues in African Theology

J. Kwabena Asamoah-Gyadu PhD

Trinity Theological Seminary, Legon, Ghana
kwabena.asamoahgyadu@gmail.com

"The earth is the Lord's and all that is in it," are the words with which Psalm 24 begins.[1] We have the same thought in Deuteronomy 10:14, where it is declared that "the heaven and the heaven of heavens belong to the Lord your God, the earth with all that is in it." In that text, divine ownership of the earth is linked to the choice of Israel as God's covenant people and underlying the promise of salvation was the bequeathing of land to them. The salvation of God's people is discussed within the context of land ownership, with implications for care and thanksgiving. Although these texts come from the Hebrew Scriptures, the theological understanding of the created order in terms of divine wisdom and activity is not limited to that civilization with its ontological orientation. These biblical and theological texts and what this book affirm are contrary to evolutionary theory as propagated by Darwin.

The Call to Ecological Theology

This book, which is on the Bible and ecotheology in Africa, specifically from the Ghanaian and Burkina Faso contexts, is an important one because it is written from an African context that shares and at the same time bridges the ontological worldviews of traditional African and Jewish societies regarding environmental care. The ecological systems in the Jewish and African contexts are so tied up with divine activity and interest

1. All Bible quotations are from the *New Revised Standard Version*.

that in the Bible, the beginnings of the story of human salvation is set within a garden, Eden. Theologically speaking, in both the African and Jewish contexts, it is impossible to depend on the environment without acknowledging the importance of transcendent beings. Not only that, but when God granted the human beings "dominion" over the created order, an important implication was the fact that our survival and very destiny depended on exercising care over the environment because the very future of human life depended on such action. Thus, Israel brings its first fruits from the land to Yahweh in gratitude and in African thought systems, and festivals are occasions of thanksgiving to gods and ancestors for sustenance made possible by eking out a living from nature.

These thoughts resonate with African worldviews because the custodianship of the ecosystems, it is believed, lies in the hands of transcendent beings. Human beings are only custodians into whose care the environment has been entrusted. If in Israel the land belonged to Yahweh, in African traditions, it is given by the Supreme Being and entrusted to humanity by the ancestors who previously served as its custodians. On the biblical side of things, Christopher J. H. Wright notes as follows:

> This reciprocal relationship between God and people generated significant rights and responsibilities in the economic sphere in relation to Israel's existence on the land. Divine ownership and divine gift—these are the two fundamental theological assertions that governed Israel's understanding of their land.[2]

When thought of this way, one wonders why ecotheology has never been part of theological thinking and teaching in Africa all this time. Ecological concern is high on the theological agenda of many Western institutions today. This is due to the terrible effects of massive pollution, ozone depletion, carbon emissions, and the general ghastly negative effects of human activity on the environment. The fight over land comes next to religion in terms of conflicts within a lot of societies today. Given the rates of environmental destruction on the continent, this is a matter that ought to be given attention in the curriculum of theological education because of the influence that Christian leadership has on African society. This volume seeks to promote such an agenda.

The natural world has largely been ignored as a subject of theological education and in Christian spirituality. In the light of that experience, which should not have been the case, the underlying thesis of the biblical scholarship in this volume is that the Bible has much to say about ecotheology and

---

2. Wright, *Old Testament Ethics*, 104.

that continued desacralization would spell doom for the whole of humanity. The treasure of nature, one scholar puts it succinctly, is "under mortal threat." This threat is occasioned by human practices of unbridled reproduction, overconsumption, and the exploitative use of natural resources.[3] The discussion of environmental degradation in terms of desacralization in several of the essays shows that it is impossible to separate the natural world from its supernatural influences in the African imagination. Africans live in a world that is filled with numberless spirits that influence life either for good or for ill. African peoples still maintain an understanding of things in which no human activity, including their economic lives, may be separated from their religious practices.

The factors that have affected the economic landscape in Africa and that have changed its face and socioeconomic orientation include colonization, the impact of religious missions and religious pluralism, the globalization of cultures, migration, technological change, and the media in all its forms. Discussions about environmental care and desacralization in Africa usually overlook the impact of capitalism and modernity on Africans. In capitalism, profits remain the ultimate goal and therefore environmental exploitation in search of mineral wealth, harvesting of precious trees, and the destruction of river bodies proceeds without much thought being given to the implications of these human activities for our future.

African Cosmologies, Environment, and Economic Life

Traditionally, the inseparability between religion and life in African philosophical thought means it is virtually impossible to talk about morality/ethics on the one hand and African economic life on the other without reference to spirituality or lived faith.[4] Religion is pervasive in African ways of life, culture, and being. Religion brings together the asymmetrical worlds of the seen and unseen, visible and invisible, or physical and spiritual realms in relationships of constant and perpetual interaction. The abundance of rain, food, fish, and other resources for sustenance all depend on the sorts of relationships that the living have with the world of spirits. The environment is the natural habitat of the spiritual beings which are considered powerful. In traditional or primal economies, when the spirits are angry because they have been neglected by the living, what results is environmental and climatic chaos, including thunder, floods, and drought. In the African traditional

---

3. Johnson, "Losing and Finding Creation," 4.
4. Magesa, *African Religion*, 3.

context, traditional African societies still maintain cordial relationships with the supernatural worlds of spirits and ancestors through pragmatic environmental practices in order to avoid these calamitous situations.

In view of the ontological worldview, African traditional economies have always been foregrounded in the cosmological idea that we live in a "sacramental universe" in which the physical serves as a vehicle for the spiritual. In the traditional thought of the Nuer of Sudan, for example, spirits of the above are usually associated with the sun, rain, lightning, thunder, and so on, whilst their spirits of the below may be identified with rivers, streams, lakes, land, etc. The intersection of the activities of the spirits of the above and below on traditional economies lies in the fact that it is the sun and rain that make it possible for the land to retain its fertility and produce food for the sustenance of human beings and their communities. The best way to conduct life is therefore governed by the relationship between human beings and the world of spirits. The breach of such simple traditional ethical codes as not fishing or farming on certain sacred days, it is believed, could affect the productivity of the land and rivers. In other words, famine and hunger could be seen as religious issues in African societies. The preparation of the land, the sowing of the rice, and its harvesting among the traditional Mende of Sierra Leone, for example, all involve extensive rituals aimed at acknowledging our reliance on nature for economic wellbeing.[5]

In the African traditional context, the relationship between spirit/religion on the one hand and ethics and economics on the other may therefore be best appreciated from the perspective of a key theme in African ontology. The natural sources of economic activity and of living and survival, like the land, forests, rivers, and even the sea, are all deified. When certain rivers dry up, the traditional explanation would usually be that the spirit that inhabits that particular natural environment may have departed, in which case it would be up to the those with the powers of divination to explain the reasons for the turn of events to the community and provide directions for religious rectification. These natural resources, as we would like to refer to them within modern civilizations, are therefore supposed to be handled with the utmost reverence and care. In *Theology in Africa*, Kwesi A. Dickson points out the description of the gods of Africa as "nature gods" must be understood to mean that the various aspects of nature are held to be the means whereby reality is experienced: the stone, the sea, the tree, and generally the various elements in the human environment, are meaningful to the African because they point to something beyond

---

5. Little, "Mende in Sierra Leone," 118–19.

themselves.⁶ The human being and communities are in kinship with nature because humanity is sustained by nature's bounty.

Eking Out a Living from Nature

What this means is that in African traditions, nature or the environment ought to be treated with ethical and moral respect because, firstly, it belongs to the now dead, but spiritually living cult of ancestors; secondly, the living are the custodians of it, holding it in trust for their living-dead ancestors; and thirdly, human life, survival, and destiny depend on it. We eke out our living from the natural resources of Mother Earth. If the African perceptions of the universe consist of the interactions of various vital forces—physical and spiritual, seen and unseen, earthly and heavenly—then the African ethical consciousness cannot but be a religious one.⁷ On account of the dynamic relationship between sacred and secular realities, Laurenti Magesa notes how the world becomes even more important in African ethical thought because it is recognized that human life depends directly upon its vital forces.⁸ Human beings eke out their living from natural resources that are owned by the Supreme Being and that are inhabited by the presence of deities and which the ancestors have bequeathed to the living.

Until fairly recently when urbanization threw its peculiar commercial challenges at Africans, land was only leased for temporary use, not perpetually commercialized for individual economic profit. Traditional chiefs and elders constantly bemoan the fact that when things worked properly with due reverence for the ancestors, no human being touched a tree, rock, or the land without first deferring to the controlling spiritual powers who ultimately are the custodians of the environment on which life depends. Economically speaking, human flourishing in traditional thought was impossible to explain apart from our relationship to, and consequently a proper and reverential handling of, the earth and its fullness as belonging to the Supreme Being and the supernatural world in its various expressions, whether as deities, spirits, or ancestors. In African ethical understanding, the earth is given to humanity as a gratuitous gift and all human beings possess an equal claim to it and the resources it offers. What this means is that any individual can only hold land in trust for one's descendants on behalf of the clan or ethnic group. Water sources, mineral resources, and forests are,

---

6. Dickson, *Theology in Africa*, 161.
7. Magesa, *African Religion*, 57.
8. Magesa, *African Religion*, 60.

in principle, public property. They are communally owned and have to be cared for and used as such.⁹

One way to appreciate the relationship between the resources of nature on which human life and well-being depends and the supernatural forces of beneficence is to look at the religious nature of the celebration of festivals. In most African cultures traditional or ancestral festivals are related to how people eke out their living, thus creating and sustaining a relationship between these ritual celebrations and economic life. Libation prayer is the way in which humans communicate with the world of spirits. The officiant at a typical libation prayer among the Akan peoples of Ghana, for instance, first invokes the forces of beneficence beginning with the Creator God, followed by Mother Earth, the ancestors, and the pantheon of lesser deities in that order. What such libation prayer petitions the world of ancestors for includes good health and agricultural abundance for feeding the living.

Mother Earth

The earth is gendered feminine among the Akan of Ghana and is the center of many a traditional celebration because she is the one who mothers the community by providing the means of economic sustenance. Among the Gikuyu of Kenya the earth is the "mother of the tribe"; it is the soil that feeds the child through its lifetime and so the earth is the most sacred thing above all that dwell in or on it.¹⁰ In most communities therefore, Mother Earth has sacred days on which no tilling is to be done on it. This is to avoid desacralization through overexploitation. It is also forbidden to desacralize the earth through the spilling of innocent human blood and engaging in sexual intercourse on the bare ground. There are trees in African forests that cannot be felled, even for economic reasons, without asking permission from the earth through prescribed rites of libation and rituals and sacrifices. The earth has spiritual power, Kofi A. Busia argued, because it is her spirit that makes the plants grow; she has the power of fertility and offerings are made to her not only so that she can help crops grow, but also so farmers may be protected from misfortune.¹¹

The invocation of the beneficent powers of Mother Earth during libation prayer is instructive because it emphasizes the point that the main source of human livelihood is sacred. A typical libation prayer among the Akan of Ghana requests from the supernatural realm of the Supreme Being,

9. Magesa, *African Religion*, 61.
10. Kenyatta, *Facing Mount Kenya*, 21.
11. Busia, "Ashanti," 195.

ancestors, and deities—seen as the sources of power, fruitfulness, and vitality—such things as make for a good life:

> God, the dependable, drinks
> *Asaase Yaa,* [Sacred Earth], drinks
> As you receive drinks,
> We seek life and prosperity.
> We seek long life,
> Business prosperity, love within the lineage.
> Drive the evil one far beyond.
> To our health, those here assembled;
> To the health of our gods and our souls![12]

All of life's forces in African traditions are intended to serve and enhance the life force of the human person and society.[13]

In the relationship between spirit/religion and economies in the traditional context, human behavior is important for its ontological consequences, which could be for good or for ill. In Chinua Achebe's classic novel *Things Fall Apart*, we read about the Igbos' respect for ancestors as depicting the kind of harmonious relationship that must exist between humans and the environment.[14] The linkage of the economic prosperity of Okonkwo, the novel's main character, with his religion, is instructive. When Okonkwo, in a fit of rage, defied traditional ethics and religious and social norms to physically assault his wife on a sacred day, the consequences of his action were interpreted in the light of its effects on the wider society. The woman had gone to plait her hair at a neighbor's home, as a result of which Okonkwo's evening meal was delayed. On return she received heavy beating from Okonkwo, who in his anger had forgotten that this was the week of peace or sacred week. This is usually a period of silence imposed on traditional societies in which people cease from working the land, fishing, or engaging in any economic activity.

Achebe writes that it was unheard of for one to beat another during the sacred week. At dusk, Ezeani the priest of the earth goddess Ani called on Okonkwo in his *obi* or hut. Against the backdrop of customary courtesy towards visitors, Okonkwo had brought out kola nut and placed it before the priest in observance of the traditional mode of welcoming strangers. The rest of the story is best told in Achebe's own words:

12. Yankah, *Speaking for the Chief*, 176–77.
13. Magesa, *African Religion*, 51.
14. Achebe, *Things Fall Apart*, 10.

> Take away your kola nut. I shall not eat in the house of a man who has no respect for our gods and ancestors . . . Listen to me . . . you are not a stranger in Umuofia. You know as well as I do that our forefathers ordained that before we plant any crops in the earth we should observe a week in which a man does not say a harsh word to his neighbor. We live in peace with our fellows to honor our great goddess of the earth without whose blessing our crops will not grow . . . The evil you have done can ruin the whole clan. The earth goddess whom you have insulted may refuse to give us her increase, and we shall all perish.[15]

The concluding words of the priest drew out the religious implications of the actions of Okonkwo. The actions of a single individual in breaking what the forefathers had ordained had the potential to affect the economic fortunes of a whole community: "the earth goddess whom you have insulted may refuse to give us her increase, and we shall all perish."[16]

Implications: African Worldviews, the Environment, and African Theology

The essays in this volume come to complement some important ones done within the Western context, but most importantly they demonstrate the growing importance of the environment in biblical and theological discourses. It is very possible to discern the importance of the created order in theological reflection, including Christology. To cite Elizabeth A. Johnson:

> The Christian scriptures, while focused on life in Christ, bear rich themes that are implicitly earth-affirming: incarnation, where the Word becomes flesh and so enters into the living matter of this world; resurrection of the body, signifying an eternal worth to the flesh; eucharistic sharing, in which bread and wine made from grain and grape usher the assembled people into communion with the Divine; and hope, that the future will bring on a cosmic scale the gracious redemption that has already occurred in Christ, the firstborn of all creation.[17]

Given the importance of the environment in African philosophical thought systems and against the backdrop of its desacralization, it should be impossible to study theology in Africa without due attention to the

---

15. Achebe, *Things Fall Apart*, 30.
16. See Asamoah-Gyadu, "Evil You Have Done," 46–62.
17. Johnson, "Losing and Finding Creation," 5.

environment. The chapters range from studies on the environment from biblical to theological perspectives, with quite a few even looking specifically at cases of degradation in Ghana. The African ontology is very much a religious one, as we have discussed, and its implications for the traditional economy are taken very seriously. One reason why the African might be more inclined towards the Old Testament, Kwesi A. Dickson argued, for example, is that in Africa, as in ancient Israel, religion pervades life.[18] African Christians find in the Old Testament a kindred atmosphere because the latter contains bodies of rules and regulations which were meant to govern the ritual and moral lives of the ancient Israelites.[19]

Many of these relate to how communities eke out a living from natural resources. Agricultural pursuits, Dickson noted, bring the African face to face not only with God, who brings rain and sunshine, but also with the Goddess of the earth.[20] The resonances between the traditional forms of faith and ecotheology could not have been better discussed apart from the essays in this volume. This calls us to engage in a new African theology of salvation that not only encapsulates salvation from sin with its eschatological implications, but which also takes the redemption of our ecosystems seriously. This is an approach to salvation that is evident in the epistle of Paul to the Romans:

> For the creation waits with eager longing for the revealing of the children of God; for the creation was subjected to futility, not of its own will but by the will of the one who subjected it, in hope that the creation itself will be set free from its bondage to the decay and will obtain the freedom of the glory of the children of God. We know that the whole creation has been groaning in labor pains until now; and not only the creation, but we ourselves, who have the first fruits of the Spirit, groan inwardly while we wait for adoption, the redemption of our bodies (Romans 8:19–23).

In other words, both creation and humanity groan for existential and eschatological redemption. We cannot talk about salvation without being holistic in our thinking. That is the stark lesson that we must take from the essays in this volume. The implications for this unitary perception of the salvific order for ethics is not too hard to find. In the Christian message, change begins with accepting Christ as one's Lord and personal Savior. In reality then, we cannot talk about prosperity without commitment to doing

---

18. Dickson, *Theology in Africa*, 154.
19. Dickson, *Theology in Africa*, 153.
20. Dickson, *Theology in Africa*, 156.

justice, loving kindness, and walking humbly with our God (Mic 6:8). And this includes a responsible attitude towards the created order entrusted to our care by God.

Bibliography

Achebe, Chinua. *Things Fall Apart*. London: Heineman, 1958.
Asamoah-Gyadu, J. Kwabena. "The Evil You Have Done Can Ruin the Whole Clan: African Cosmology, Community, and Christianity in Achebe's Things Fall Apart." *Studies in World Christianity* 16.1 (2010) 46–62.
Busia, Kofi A. "The Ashanti." In *African Worlds: Studies in the Cosmological Ideas and Social Values of African Peoples*, edited by Daryll Forde, 190–209. Oxford: Oxford University Press, 1954.
Dickson, Kwesi A. *Theology in Africa*. Maryknoll, NY: Orbis, 1984.
Johnson, Elizabeth A. "Losing and Finding Creation in the Christian Tradition." In *Christianity and Ecology: Seeking the Well-Being of the Earth and Humans*, edited by Dieter T. Hessel and Rosemary Radford Ruether, 3–21. Cambridge, MA: Harvard University Press, 2000.
Kenyatta, Jomo. *Facing Mount Kenya*. London: Mercury, 1938.
Little, Kenneth. "The Mende in Sierra Leone." In *African Worlds: Studies in the Cosmological Ideas and Social Values of African Peoples*, edited by Daryll Forde, 111–37. Oxford: Oxford University Press, 1954.
Magesa, Laurenti. *African Religion: The Moral Traditions of Abundant Life*. Maryknoll, NY: Orbis, 1997.
Wright, Christopher J. *Old Testament Ethics for the People of God*. Leicester: InterVarsity, 2014.
Yankah, Kwesi. *Speaking for the Chief: Ókyeame and the Politics of Akan Royal Oratory*. Indianapolis: Indiana University Press, 1995.

Introduction

The Bible and Caring for the Land

Benjamin Abotchie Ntreh

Abstract

This paper tries to read selected biblical texts to address the topic. I did my reading as a Ga of Ghana, in as much as I share the ethnic worldview of seeing the land as composed of three facets: earth, sky and sea (water bodies). I read from the creation narratives in the book of Genesis to address the issue of care for the land. I have shown in this paper that care for the land is care for humanity, contrary to the predominant notion that had existed for a long time that humans are the center of everything; a notion that gave rise to the destruction of the land. I have also shown that open defecation and disposal of solid waste into our water-bodies are not consistent with biblical records. The biblical writers had cautioned ancient Israelites to "dig," "drop," and "cover" their feces. I have also talked about protection of our water-bodies, as a protection for ourselves.

Introduction

To a very large extent Christians have compartmentalized their lives. They read the Bible in church and sometimes in their homes, but they live their lives separate from what is read from the Bible. Although the majority claim to be adherents of the Bible—the word of God for life—the teaching and tenets of the Bible are not seen in their lifestyles. This is the reason for my call to make our voices accessible to the church so that we can influence the lifestyles of members of the church about the relationship with land.

Let no one think that Africans are the only people who have a problem caring for the land. The Western world also, until recently, had the same problem. In his article entitled "Ecological Approaches: The Bible and the Land," Gene M. Tucker has this to say, "The Green Bible is the definitive movement Bible that shows that God is green and how we can care for and protect God's creation."[1] On the part of Africans, I will say that we seem to have retrogressed from the way our forebears revered the land.

This paper will touch on some biblical texts that address issues that pertain to caring for the land. The method I will be using is African Bible hermeneutics, which I proposed some time ago. This is a dialogical interplay between the African reader and the biblical texts on the subject matter. Here, the African reader is a scholar of the Bible as well as a flesh and blood African who shares in the successes and woes of Africa. In this particular paper, I am speaking as a Ghanaian in dialogue with the word of God as I seek to address the issue of caring for the land.

I was ashamed to hear in 2015 that one of the newspapers in Ghana, *The Daily Graphic*, "carried an article declaring Ghana the 7th dirtiest country in the world."[2] The news item also said that in West Africa, Ghana is second from the bottom. June 5 is World Environment Day. It was sad to note from a JoyFm Breakfast show on the morning of June 5, 2018, that whereas Kumasi used to be called the Garden City (it was not only the Garden City of Ghana; it was the Garden City of West Africa), it is now ranked eleventh. In fact, there is nothing "garden" about Kumasi today. I will tackle the topic of the Bible and caring for the land in this manner: (1) how the Bible and the Ghanaian view land, and how they (2) care for the land: (i) the earth, (ii) the air, and (iii) the sea and water bodies.

How the Bible and the Ghanaian View Land

It is very difficult to have a working definition of the land unless one chooses a text to work with in defining what the view of the Bible is concerning the land. Our first encounter with land in the Hebrew Scriptures is in the creation account in Genesis 1:1. The Hebrew word *'ereṣ* (Gen 1:1) is translated principally as "earth" or "land." In his work, *The Land*, Walter Brueggemann has pointed out that in the Hebrew Scriptures, land is always used as a symbol. It is not spoken of simply as the physical, tangible, and solid mass on which humans live. Rather, land is always historical. Ancient Israel is closely

---

1. Tucker, "Ecological Approaches," 349.

2. Wijesekera, "What Does it Really Mean?," para. 1. https://www.unicef.org/ghana/media_9678.html

related to the land and what happens to them in the land.³ Among traditional Ga folks, there seems to be a trinity of the land—the *sky*, the *land,* and the *sea*. I will look at the biblical view of the care of the land from these three positions. It is clear from the biblical account that God gave responsibility to the humans he had created to have "dominion over the fish of the sea, over the birds of the air and over the cattle over all the earth and over every creeping thing that creeps on the earth" (Gen 1:26). It is clear that in the Ga and Hebrew cosmologies, land constitutes earth, the sky (air), and the sea (water bodies). In the creation accounts, the earth (the land) and the sea are strictly the abode of some creatures of God, but the sky (the heavens) is largely the abode of God, although birds fly in the sky. I am going to restrict myself to these three constituent components of the land in addressing the care that we as human beings must give to the land.

Perceptions about Caring for the Earth

Dominion:

It is important from the onset to mention that the use of the word "dominion" in Genesis is not in the typical destructive sense. Another Hebrew word, *radah,* which has been translated as "subdue" in the English language, appears alongside "dominion" and does not bear the notion of destruction. This means that *radah* and *kabash* are synonymous. I am reminded that the destructive sense that has been attached to the word "dominion" (*radah*) in Genesis 1:26 was a very late phenomenon. Claus Westermann argues that these two words most probably emerged from the role of the king. However, as the Ga would say, "It is the people who install a king."⁴ That means that without a people there can be no king. Thus, it is the responsibility of the king to make sure that the people are satisfied for him to remain a king. Exercising dominion over one's own people and subduing them is culturally unacceptable. Yet a notion of the destructive has been used to interpret these words in the Genesis text. This does not seem to be the original meaning of the Hebrew Scripture. Writing on this late destructive phenomenon, Richard Bauckham has this to say, "It was Francis Bacon, in the seventeenth century, who hijacked the Genesis text to authorise the project of scientific knowledge and technological exploitation whose excesses have given us the ecological crisis."⁵ Claus Westermann

---

3. Brueggemann, *Land*, 3.
4. Westermann, *Genesis 1-11*.
5. Bauckham, *Bible and Ecology*, 6.

has directed us to more plausible meanings of *radah* and *kabash* in Genesis 1. He urges us to read Genesis 1:26 in the light of Genesis 1:16 (*mashal*, "to rule") and 1:29 (where leafy green plants were to be food for the humans). He writes, "The conclusion from the latter is that dominion over animals cannot mean killing for food."[6] He continues:

> It is here that we must look to 1:16. We saw that there "to rule" can only have a nonliteral meaning—the sun rules over the day, the moon over the night; that same meaning is appropriate to 1:26b; among living beings, humans rule over the animals without condition. It is quite possible that we have here an echo of the belief that the animal was the human's deadly enemy in the early stages of the human race, and that consequently the person's dominating role in relation to the animal is saying something that concerns our very existence. Dominion over the animals certainly does not mean their exploitation by humans.[7]

This same notion can be extended to the land itself. In Genesis 1:28, humans are to "fill the earth, and subdue *(kabash)* it." Here, it is a reference to the earth itself that is to be subdued. However, it is difficult to see any destructive notion here when the import of the blessing is for humans to fill the earth. Talking about the relationship between *kabash* here and *radah*, used earlier with reference to the activity of humans, Westermann writes: "It is possible that this verb too derives from the rule of the king, even though it is not used in this context in the Old Testament. In practice it has the same meaning a רדה" (*radah*).[8] Having set the stage for this context, we shall take a closer look at the three constituent parts of land—the earth, the sky (air), and the sea (water bodies) as they relate to the care of the land by human beings.

The Earth (Land)

It seems God created us and put us here on earth to find what His will is for us, for our own good. I remember that one of the theories I learned in secondary school about creation was that matter is neither created nor destroyed. Now we know that it is not true. Knowledge is dynamic. It is a well-known fact today that matter can be destroyed. In the creation narrative in Genesis 1, God called the earth (land) into being and the first duty he gave to the human

---

6. Westermann, *Genesis 1-11*, 159.
7. Westermann, *Genesis 1-11*, 159.
8. Westermann, *Genesis 1-11*, 161.

being was to have dominion over created things, including *the earth* (Gen 1:26). Our care for the land is intricately related to our care for the other living creatures. As I have said above, to have dominion over something does not mean to exploit or destroy that thing. Rather, today many scholars "have rightly seen in Genesis 1:26 the implication that human dominion is some sort of reflection of God's rule, and therefore that in some sense the Bible's portrayal of God's rule should be the model for humanity's."[9] Our actions in relation to the land and the created world is on behalf of God, who has absolute control over all things. Thus, humans need to ask, at all times, whether God would do it the way we want to do it.

The next duty God gives to humanity is to fill the earth and subdue it (Gen 1:28a). I have already pointed out in my definitions above that the word *kabash*—translated as "subdue"—does not mean the destruction of the earth. Clearly here in Genesis 1:28 *kabash* is placed in line with being fruitful, multiplying, and fill the earth. These are positive qualities expected of human beings. The words of Richard J. Clifford and Roland E. Murphy on the expression "to subdue the earth" are very revealing. They write, "The nuance of the verb is 'to master,' 'to bring forcefully under control.' Force is necessary at the beginning to make the untamed land serve humans. Humans nonetheless are to respect the environment; they are not to kill for food but are to treat all life with respect."[10] Although Clifford and Murphy talk this way, they do not seem to consider the earth as a living entity. I contend that the subject of the verb "subdue" is the earth itself. I also contend that the negative notions ascribed to how humans should treat the earth are absent in the mind of the Hebrew writer of Genesis 1. Humans are expected to cultivate the land and make it fruitful for their own sustenance. It is also important to note that in the creation story itself humanity was not expected to kill the animals for food. Humanity was initially supposed to be vegetarian (Gen 1:29). Later in the book of Genesis humans were allowed to kill and eat animals. The text reads, "Every moving thing that lives shall be food for you; and as I gave you the green plants, I give you everything" (Gen 9:3). Bauckham sees this as a reformulation of the creation account. He writes: "The human dominion over other living dangerous animals (Gen 9:2), and humans are permitted to kill and eat animals, with the proviso that the sacredness of all life be acknowledged by abstention from blood (Gen 9:3-4)."[11] It is important to note that even here humanity was not given unrestricted permission to kill at will and eat all living creatures.

---

9. Bauckham, *Bible and Ecology*, 31.
10. Clifford and Murphy, "Genesis," 8–43.
11. Bauckham, *Bible and Ecology*, 119.

Elsewhere, in a paper called "The Survival of Earth: An African Reading of Psalm 104" that was published in *The Earth Story in the Psalms and the Prophets*, I had argued that God expects us as humans to live in harmony with other created things.[12] My conclusion in that paper is worth reiterating here:

> I have highlighted the fact that the predominant traditional views of African peoples concerning Earth are positive and reflect the principles of mutual custodianship: when we take care of Earth it takes care of us. That principle also seems to be reflected in Psalm 104 where we discover God using Earth to care for us. However, 'civilization,' together with [a] few African sayings that promote the belief that human life is the ultimate and the most essential reality on earth, have distorted the traditional norms and values of African peoples toward earth, with the result that earth is continually being destroyed. And the danger is that if the trend continues, life on earth will become impossible. I contend, therefore, that Psalm 104 calls us back to the wisdom of our forebears, to learn and to apply their values toward earth—both for its survival and our own.[13]

If we shake that equilibrium, we do so at our own peril. In this direction, animals are allowed to operate in their domain and humans do not encroach on the space of animals. In this way humans and animals can live in harmony. Disruption of this harmony is disastrous for both animals and humans. We as humans create a problem for ourselves when we think that we are the only living things that matter. We are not autonomous. We are joint creatures with the rest of creation. Bauckham has argued that Jesus depended on Psalm 104 when he preached the Sermon on the Mount in Matthew 6:33. However, to make this connection, it is important to hear what Benedict T. Viviano has to say about what Jesus means in Matthew 6:33. He writes, "The ultimate goal of all our activity must be the highest value, the kingdom of God, which is defined here as justice." He continued, "The justice envisaged is not justice in God alone but one that we are to produce on earth ourselves."[14] On Jesus' usage of Psalm 104, Bauckham writes:

> Both Psalm 104 and Jesus challenge us with the conviction that the God-given resources of creation are sufficient for all God's creatures—that is, for the reasonable needs of all God's creatures, not for the kind of excess in which, of all God's creatures,

---

12. Ntreh, "Survival of Earth," 98–108.
13. Ntreh, "Survival of Earth," 108.
14. Viviano, "Gospel According to Matthew," 630–74.

only humans indulge. God's provision is sufficient if equitably shared.[15]

Yet care for the land is not limited to treating animals well. Care for the land requires our careful treatment of the land itself. It is a very sad spectacle to see people buying wetlands and spending so much money to reclaim the land for putting up a building. Such people are aware that the land is a wetland, yet they try very hard to subdue that piece of land. For a while such people seem to be in control until rains come and floods erode the foundations of the building. When we are able to reclaim the wetlands and floods do not disturb us, we lose the water level in the land for boreholes and other things, including farming. Care for the land is care for our own survival.

Elsewhere in the Hebrew Scripture there are passages that suggest how we should take care of the land itself. In the account of the fall of humanity in Genesis 3:23, human beings are sent out of Eden to go and work the land. On this issue Bauckham writes, "Later he is sent to perform the same task outside Eden: 'to till the ground from which he was taken' (3:23)." Again, he says, "The man from the soil must work the soil in order to live from the soil's produce."[16] This may very well be in line with the earlier injunction given to humans in Genesis 1:26 and 28. Later in Genesis 9:20, however, the details are given as to what humans are supposed to do. It is here that we are told Noah "began to be a farmer, and he planted a vineyard." Incidentally, Noah is said to be the first human to cultivate the land. This seems to be putting into reality the injunction to Adam "to till the land" in the second creation account (cf. Gen 2:5). This act of Noah is said to bring joy and relaxation. In the words of Westermann, "Over and above the toil and labour of the farmer to produce the necessities of life, it yields a product that brings joy and relaxation."[17] It is important to note that in ancient Israel sitting under one's own oak trees and/or vineyard and drinking the produce of one's own vine are seen as the ultimate blessing humans can obtain from God (cf. 1 Kgs 4:25; Mic 4:4; Zech 3:10). How can we receive the blessings of God when we destroy the source of those blessings, that is the earth, the land?

There used to be a slogan on Ghana Broadcasting Corporation's GTV on the environment that goes like this: "When the last tree dies, the last man dies." In Deuteronomy 20:19–20, we are given some regulations that must be observed even in times of war. The text reads:

---

15. Bauckham, *Bible and Ecology*, 75.
16. Bauckham, *Bible and Ecology*, 22.
17. Westermann, *Genesis 1-11*, 487.

> When you besiege a city for a long time, while making war against it to take it, you shall not destroy its trees by wielding an axe against them; if you can eat them, do not cut them down to use in the siege, for the tree of the field is man's food. Only the trees which you know are not trees for food you may destroy and cut down, to build siege-works against the city that makes war with you, until it is subdued.

It is assumed that during wartime soldiers can freely become disastrous and treat their enemies as well as the environment with disdain. Here in Deuteronomy 20:19–20, soldiers are entreated to be caring even for the trees in the city around which they have laid siege. Food-growing trees are supposed to be saved. Even nonfood-producing trees may only be cut to build siege-works, otherwise soldiers have no business cutting down trees during wars (cf., 2 Kgs 3:19). I see a certain similarity between this requirement and how in our traditional settings people are not allowed to wantonly cut down food crops when they take possession of lands. People are urged to wait till the food crops are harvested, even by encroachers, before they take possession of lands.

Open Defecation & Air (Sky) Pollution:

The introduction to an article on MyjoyOnline, a news item from August 21, 2017 on sanitation, reads:

> Ghana is in [a] dire state when it comes to sanitation. One out of every five Ghanaians defecates in the open and 90% of all our excreta end up in streams and rivers—the same place from which we get our potable water. If Accra, the national capital, had to dump human excreta into the sea for several years, then one can only imagine what happens in other towns and cities across the country. Solid waste management, on the other hand, is nothing to write home about although private sector involvement has caused some improvement.[18]

This is shameful about Ghana, a country that claims to be "the gateway to Africa" in the second decade of the twenty-first century. The place where human excreta is poured into the sea in Accra is sarcastically called "Lavender Hill" due to the pungent smell that engulfs the entire environment. Open defecation pollutes the air (sky). However, a little over 2,400 years ago,

---

18. https://www.myjoyonline.com/news/2017/august-21st/front-pages-monday-august-21-2017.php

the writer of Deuteronomy addressed the issue of open defecation. Deuteronomy 23:12–14 exhorted ancient Israelites against open defecation. On their journey to the promised land, ancient Israel was instructed to have a place outside their camp where they may go and evacuate their bowels. At this place they were told, "You shall have an implement among your equipment, and when you sit down outside, you shall dig with it and turn and cover your refuse" (Deut 23:13). The reason for this instruction is that God abhorred uncleanliness when he came among his people (v. 14). Incidentally, God put this instruction into the makeup of cats. When cats are to ease themselves, they first dig the ground, ease themselves into the hole and they cover it up. It is sad that in twenty-first-century Ghana, people defecate in the open and they leave their feces uncovered. Can we claim to be better than cats? I doubt it. Open defecation does not only degrade the earth—the environment—it pollutes the air.

Genesis 8:21 seems to be cast in the mode or worldview of nations of the ancient Near East, where the gods smelled the aroma of human sacrifices. In this direction, Clifford and Murphy write, "Yahweh smells the pleasing odour and promises never to repeat the universal punishment."[19] Although this is an anthropomorphic representation of God and how God operates, it tells us that Yahweh accepts the sacrifice of Noah. Westermann has pointed out that this notion of Yahweh cast in human terms as Yahwistic persisted to the latest periods (Exod 29:18, 25, 41; Lev 1:9; Ezek 6:13).[20] Conversely, the pungent smell from our pollution of the air also becomes an affront to God. God would reject sacrifices or any burnings that were not pleasing to him, just as he did to sacrifices of cities that did not please Him (cf. Lev 26:31). Whatever we do on earth affects the sky (the air). God smells the sweet aroma of human sacrifices and accepts them. In the same manner, the pollution that we make pollutes the sky and thus becomes abhorrent to God. It is strange to know what has happened to our psyche. The groves and environments of our shrines are kept clean. The forest was believed to be the abode of the deities. How come we now desecrate the earth with wanton abandon? Are we doing any better with our care for water bodies?

Sea (water bodies)

It was common some time ago to hear on television "Water is Life." Our forebears forbade the destruction of wetlands. Wetlands were sources of fish and potable water for most of our communities. When this was the case,

19. Clifford and Murphy, "Genesis," 16.
20. Westermann, *Genesis 1-11*, 454.

wetlands were kept clean and some were recognized as sacred places. This is because our forebears recognized that water helps to sustain life. However, with the introduction of pipe-borne water, the significance of the wetlands has diminished. Thus, in recent times it is a common sight to find people using so much money to fill wetlands for development. A decade ago, when people started suffering from burilli ulcers, I knew there must be something wrong with our water bodies—the sources of drinking water for most of the sufferers. A Joint Monitoring Program for Water Supply and Sanitation of UNICEF and WHO report released in 2015 gave the population ratio between the urban and rural population as 51 percent and 49 percent, respectively. However, access to basic water was 88 percent for the urban population and 66 percent for the rural, leading to an average of 78 percent. Yet the percentage of people with connected potable water to their houses: was 33 percent for the urban population and 3 percent for the rural population—an average of 18 percent.[21]

In Leviticus 11:36, springs and cisterns are considered legally pure places and things. Water is a source of purity (cf. Num 19:17–22 Zech 13:1). In an article presented at a UCC-UNILORIN Conference in Illorin, Nigeria, I argued that there is human complicity in the flood narrative in Genesis 6:5—8:22.[22] Similarly, in 2015, Jonathan Kavusa Kivatsi presented a PhD thesis entitled "The Life-Giving and Life-Threatening Potential of Water and Water Related Phenomena in the Old Testament Wisdom Literature" to the University of South Africa (UNISA). From both the canonical and extracanonical wisdom books of the Hebrew Scriptures, Kivatsi shows that water and water-related phenomena have the potential to give and destroy life. When handled properly and unimpeded, water can be a source that gives life to humans. However, if humans impede the waterways, whether through diversions, reclamations, or by narrowing them, water can threaten human life. Most of the floods in our cities are a result of human actions and inactions.[23]

There are few references from the New Testament to address this issue of land care. This may be due to the subject matter and interest of most of the New Testament writers. However, the New Testament writers were not oblivious to this issue. I have already mentioned an instance of Jesus' position earlier. Again, a line in the Lord's Prayer is important here: "Your

---

21. "Water Supply and Sanitation in Ghana," para. 1.
22. Ntreh, "Human Complicity in Climate Change," 91–102.
23. Kivatsi, "Life-Giving and Life-Threatening Potential."

will be done *on earth as it is in heaven*" (Matt 6:10b; Luke 11:2b).²⁴ On this, Bauckham writes:

> The coupling of 'heaven' and 'earth' cannot fail to evoke the whole of creation, everything God created at the beginning (Gen 1:1; 2:1 and 4). God, it was standardly said, is the Creator of Heaven and Earth, and this is the basis on which his Kingdom must come on earth as it is in heaven. The Kingdom does not come in order to extract people from the rest of creation, but to renew the whole creation in accordance with God's perfect will for it.²⁵

This is similar to the view expressed earlier that humans are co-creatures with the rest of the created things. God is interested in all of creation. Thus, humans must not exalt themselves above the other creatures.

Very little is seen in the Pauline literature on land care. However, in Roman 8:18–23, Paul expresses concern about the ecological degradation of his day. Things are so bad that the Earth itself mourns for a renewal. In the words of Bauckham, "Romans 8:19–23 has been described as 'an environmental mantra', meaning that appeal is often made to it as a kind of ecological proof-text, mandating environmental activity by Christians, without engaging in exegetical detail with the problems of interpreting the passage."²⁶

Conclusion

So far, we have looked at the Bible, especially the Old Testament, and our care for the land. I have shown that the Hebrew Bible uses the word "earth" and "land" interchangeably. I have looked at three components of earth—land, sky (air,) and sea—a view that is shared by the Ga of Ghana and the Hebrew. Caring for the land is all about human actions and inactions. In all three components, I have shown that care of the land is for the good of humans. In the New Testament, Jesus and Paul taught in ways that are tangential to the topic, yet their concerns border on renewal of the creation of God.

---

24. Emphasis mine.
25. Bauckham, *Bible and Ecology*, 166.
26. Bauckham, *Bible and Ecology*, 100.

Bibliography

Bauckham, Richard. *The Bible and Ecology: Rediscovering the Community of Creation.* Waco, Texas: Baylor University Press, 2010.

Brueggemann, Walter. *Genesis.* Interpretation. Atlanta: John Knox, 1982.

———. *The Land.* Philadelphia: Fortress, 1986.

Clifford, Richard J., and Roland E. Murphy. "Genesis." In the *Jerome Biblical Commentary*, edited by Raymond E. Brown et al., 8–43. Englewood Cliffs, NJ: Simon & Schuster, 1990.

Deane-Drummond, Celia, ed. *Pierre Teilhard de Chardin on People and Planet.* London: Equinox, 2006.

Ela, Jean-Marc. *African Cry.* Maryknoll, NY: Orbis, 1986.

Fox, Matthew. *Creation Spirituality.* San Francisco: Harper, 1990.

Kivatsi, Jonathan Kavusa. "The Life-Giving and Life-Threaneing Potential of Water and Water-Related Phenomena in the Old Testament Wisdom Literature: An Eco-Theological Exploration." Presented to the Department of Biblical and Ancient Studies, University of South Africa, 2015.

Ntreh, B. Abotchie. "The Survival of Earth: An African Reading of Psalm 104." In *The Earth Story in the Psalms and the Prophets.* The Earth Bible 4. Edited by Norman Habel, 98–108. Sheffield, UK: Sheffield Academic Press, 2001.

———. "Human Complicity in Climate Change: An African Biblical Hermeneutical Reading of Genesis 6:5—8:22." *Ilorin Journal of Religious Studies (IJOURELS)* 2.2 (December 2012) 91–102.

Sarna, Nahum M. *Understanding Genesis.* New York: Schocken, 1966.

Tucker, Gene M. "Ecological Approaches: The Bible and the Land." In *Method Matters: Essays on the Interpretation of the Hebrew Bible in Honor of David L. Petersen*, edited by Joel M. LeMon and Kent Harold Richards, 349–67. Atlanta: Society of Biblical Literature, 2009.

Viviano, Benedict T. "The Gospel According to Matthew." In the *Jerome Biblical Commentary*, edited by Raymond E. Brown et al., 630–74. Englewood Cliffs, NJ: Simon & Schuster, 1990.

"Water Supply and Sanitation in Ghana." https://en.wikipedia.org/wiki/Water_supply_and_sanitation_in_Ghana.

Westermann, Claus. *Genesis 1-11.* Minneapolis: Augsburg, 1984.

Wijesekera, Sanjay. "What Does it Really Mean When We Say Ghana is the 7th Dirtiest Country in the World?" https://www.unicef.org/ghana/media_9678.html.

The Bible and Environment —Care for the Land

Emmanuel Gyimah Louis Twum-Baah
All Nations University College, Koforidua

Abstract

The aim of this paper is to draw attention to the incalculable mess caused to a once-beautiful and all-sustaining environment, handed down to humans from time immemorial and of which they are called to be stewards. We argue for the need of sustained discourse on the environment. The irony is that the so-called monotheistic religions, which claim allegiance to the Tanakh as their most holy relic and the source of creation's story, do not show much concern for the destruction of the earth. The discussion re-iterates some of the negative effects of the direct human activities that are detrimental to the well-being of the environment and the land. The paper draws on popular views of people in direct and indirect discourse with the Bible, environment, and care for the land as we negotiate the curves toward an appropriate discourse of the environment. We point out that a meaningful engagement in advocacy for the integrity of creation needs no overemphasis. The urgent need for watchdog groups to challenge the status quo and guide government, traditional authorities, and the church in proactive education of the people in good land practices and relationship with creation has been pointed out. Here, WAATI has been identified as one such watchdog institution, which should not exist all for itself as an elite group, but rather rise up to champion the teaching of the Tanakh on socioeconomic, political, and environmental issues in order to save the earth.

Introduction

THE ISSUES OF THE environment resulting from direct human activities and land use in particular, can be rated amongst the most difficult subjects of the present time. The rampant wars in the Middle East and parts of Africa are disturbing and so are the natural disasters in the Americas and the Pacific region. However, the issues of environmental degradation around the world are much subtler and harder to deal with because of their political, economic, and sociocultural contexts. At the level of academic discussion too, matters become complicated by the nuances of the interpretations put on the creation story. In that situation, Christianity has been made a scapegoat for the issues of environmental degradation.[1] Others also have located the "roots of the problem in the Industrial Revolution," describing it "as the greatest enemy Nature has ever had to face."[2] Further on regarding the problems of interpretation, we are confronted with the issue of the assignation to humans to rule over all creation. We will look at later in a bit more detail later.

In time past, the Hebrew Bible (OT) was seen as a common denominator which shaped many cultures, people's worldviews, ethical value systems, and even the common law of nations. The Hebrew Bible (OT) is subscribed to by three popular world religions that claim to be monotheistic: Judaism, Christianity, and Islam. All three share "the biblical idea of God as Creator, Judge and Ruler of the universe."[3] For a long time the Hebrew Bible has been the one and only book of the world that has more written on it than any other known literary composition. According to Ernst Würthwein, "No book in the literature of the world has been so often copied, printed, translated, read, and studied as the Bible. It stands uniquely as the object of so much effort devoted to preserving it faithfully, to understanding it, and to making it understandable to others."[4]

It is without doubt that the Bible is an ancient book, but many people consider it to be ever true and relevant to life in every generation. Others do not. The text might be the same, but its acceptance over the years has changed. It is a fact of history that Western society was largely shaped

---

1. White, "Historical Roots," 1205–6.
2. Opuku-Agyeman et al., "Akan Traditional Beliefs," 141.
3. Harris. *Understanding the Bible*, 2. See also Damonte, "God, the Bible," 41.
4. Würthwein, *Text of the Old Testament*, 121.

by ideas from the Bible, but in recent times lots of those ideas have been received with skepticism. The modern scholar questions the basis of its tenets and no longer accept them "by faith" as it used to be many years ago. Indeed, the book has become anathema to many. The feminist movement is very critical of its seeming masculine posture, while advocates for same-sex relationships consider some of its teaching a trampling of their human rights. Conservationists and friends of animals and wildlife are also forces to contend with in matters of the Bible and environment. For a world that is increasingly alienating itself from the God revealed in that ancient book, the complicating interpretations are to be expected.

Notwithstanding the positions assumed by any group of people and their interpretation of the Hebrew text, it is important to have an open, independent, and objective discussion of what the Bible says concerning the environment and caring for the land.

This is how we shall proceed. First, we will review a bit of what the Bible teaches regarding the environment and then we will discuss caring for the land. This will lead on to the issues of human activity and degradation of the earth and environment. In that section we will navigate a bit into the painful issues of cities and municipalities generating waste and dumping sluice and indecomposable stuffs on rural communities. Finally, we will make suggestions regarding the way forward and thereafter conclude the discussion.

A Discourse on the Bible and Environment

The issue of environmental degradation appears on top of the world's concern list. Concerns have been expressed regarding the continued existence of humans and sustenance of the earth and everything in it, from the tiniest microscopic forms of life to endangered species near extinction, fishes of the sea and inland waters, beasts of the earth and fowls of the air. These matters have been taken up at different times and at several international forums as issues of concern to many nations. The concern about life and the sustainability of the earth evoke the creation story and intent of the Creator for it. Indeed, every effective discussion of the issues of the environment starts with the Bible's creation story. The environment cannot be discussed in isolation. The creation story in the Bible is a necessary coefficient of the issues of the state of the environment.

a. Creation of the Environment

The Creation story is recounted in the opening book of the Bible—Genesis (Heb. b*e*reshiyth, "in [the] beginning"). It is indeed "a book of beginnings, or 'origins' as its English title, Genesis (derived from the Septuagint Lev 2:4a as its likely source)"[5] suggests. Notable about the creation story is that "it begins in the distant past of creation, an event about whose absolute date we cannot even speculate, through millennium to reach Abraham at the end of chp. 11."[6] The opening sentence in Genesis 1:1 establishes that God is the sovereign of all creation. There are five words in the first verse of the Hebrew text, including *Elohim*, set between the four others to make the attribution to God.

In Hebrew morphology, "God" is the subject of the verb "created." The resultant objects of that creative action are "the heavens" (celestial) and "the earth" (terrestrial). By this sentence and because of its ramifications throughout the rest of the Bible, particularly in the Psalms, it is impossible to argue any other way who is the sovereign of all creation except to attribute it to God, the architect and owner of the objects he created. His ownership of creation is not in dispute; concerning the heavens, the earth, and everything in it (Deut 10:14), it is he who created and formed humanity (Isa 43:1). The psalmist consistently acknowledges God as Creator of the heavens and the earth. "The heavens are yours, and yours also the earth, you founded the world and all that is in it" (Ps 89:11). "The sea is his, for he made it, and his hands formed the dry land" (Ps 95:5). The finest description recorded proclaims, "the earth is the Lord's and everything in it, the world, and who live in it; for he founded it" (Ps 24:1–2a). The psalmist again observes, "I praise you because I am fearfully and wonderfully made; your works are wonderful, I know that full well" (Ps 139:14).

When God created and separated the celestial from the terrestrial, creation had no form and was empty, with darkness over the surface of the deep. The Spirit of God hovered over the waters in readiness for further instruction at the word of the Sovereign God. To bring shape and order, the Sovereign God uttered seven times his words of exhortation (Gen 1:3, 6, 9, 11, 14, 20, 24)—"Let there be"—into the chaotic creation to transform it. At the end of each execution "God saw that it was good." Thus, the environment emanates from God's creativity. It is he who divided it into the celestial and the terrestrial. Space encompasses the sky, the sun, the moon and stars, and the other heavenly bodies, some of which inhabitants

---

5. See Longman and Dillard, *Introduction to the Old Testament*, 38.
6. Longman and Dillard, *Introduction to the Old Testament*, 38.

of the earth have not yet discovered.⁷ In similar fashion, the land includes its human habitants, wildlife, the sea, and water bodies with teeming life, as well as the enclosures in the belly of the earth. After his creative activity, humans are appointed to take charge of God's creation. We shall look at the associated difficulties in the next section.

b. Assignation of humans as stewards to rule and to subdue the earth

God, in soliloquy, suggests the creation of humans "in our image, in our likeness, so that they may rule over" creation (see Gen 1:26). The mandate to rule over creation or "subdue it" (v. 28), is explained to include food for humans (v. 29) and the naming of "each living creature" (Gen 2:19). Care for the land arises out of this covenanting relationship the Creator establishes with the human he created from the dust of the earth (cf. v. 3) to share in the nurturing and preserving of all that God created. To restate the facts,

1. Humans are the only creature made in the image and likeness of God. The reason for this is that humans "may rule over every living thing God had made" (cf. 1:26, 28).

2. God created all things and saw that they were "good," then he created humanity in his image and blessed them.

3. He gave them the charge to increase in number, fill the earth, and subdue it. Part of that responsibility is to work the land and take care of it (cf. 2:15).

4. Humans are given the luxury of having "every seed-bearing plant and every tree that has fruit with seed in it" for food (1:29).

We note here that taking care of the estate of God, the land, and the created environment is a call to stewardship, which exudes awareness of a trust imposed with imperatives. In the story of creation there are cautions given to humanity in the assignation to be stewards of the earth. Humans would procreate, but also oversee the whole of creation according to rules and regulations set by the Creator. Just as the sun is appointed to govern the earth by day and the moon and stars at night (Gen 1:16–18), so are humans to preserve the integrity of all creation. This primordial covenant enjoins all

---

7. Space scientists continually report about further discoveries of new heavenly bodies. Now that NASA's Insight spacecraft has successfully landed on Mars, having done a journey of more than six months' duration and 300 miles, the world will expect to learn more about the universe (see Chang, "NASA's InSight Mission.").

humans without exception to show responsibility toward taking care of the land. This responsibility has led to various interpretations.

c. The Concerns

The fact that God chose humans to be stewards has become a contentious issue for many scholars. Even the psalmist once questioned why the "Lord, our Lord," who is "majestic in all the earth" (Ps 8:9) should entrust "care of the works of his hand to man" (Ps 8:6). "What is mankind that God should be so mindful of them and to crown them with glory and honor" (cf. vv. 4–5)? Well, the psalmist questioned the reasoning behind the Creator's gesture but later discovered that it is because "nothing is too hard for God" (cf. Ps 32:17b). It is not hard for God to give away all his estate to the care of mundane humanity.

The scholarly arguments against a traditional interpretation of Genesis 1 starts with Lynn White, who in 1967 argued in an article that Christianity is the root of the world's environmental problems. White argued that modern science was "an extrapolation of natural theology and modern technology," and was "partly to be explained as in occidental, voluntaristic realization of Christian dogma of man's transcendence of and rightful mastery over nature." Continuing the argument, White categorically declared that "somewhat over a century ago science and technology hitherto quite separate activities, joined to give mankind power which, to judge by many of the ecological defects are out of control. If so, Christianity bears a huge burden of guilt."[8]

Commenting on that article, Heather Truelove and Jeff Joireman capture the point for White's reasoning: "The crux of White's argument is his interpretation that Genesis 1 implies that humans have dominion over nature and that nature exists solely for man's use."[9] However, in their point of view and following Dobel, "Genesis 1 dictates that humans have stewardship relationship with nature and that humans should take care of and protect nature."[10] Augustine M. Mensah also rightly argues along those lines that "creation in the image and likeness of God and the injunction to rule over or dominate creation is not about exploitation but rather, nurture and care for creation, which is 'responsibility' toward God for the nonhuman creatures."[11]

---

8. White, "Historical Roots," 1206.
9. Truelove and Joireman, "Understanding the Relationship," 807.
10. Truelove and Joireman, "Understanding the Relationship," 817.
11. Mensah, "Uniqueness of Humankind," 34.

As part of their critical assessment of Christian attitude toward creation, the authors observed that "in several environmental domains Christianity has been shown to be negatively related to environmentalism." That "one consistent finding is that when environmentalism is framed in terms of a trade-off between economic interests and environmental interests, Christians, and specifically Biblical literalists, show less support for the environment than non-Christians."[12]

Marco Damonte also looks at White's concern that God created all the animals and allowed mankind to have control over them, "thus establishing his dominance over them. God planned all of this explicitly for humans' benefit and rule" and that "no item in the physical creation had any purpose save to serve man's purposes. Furthermore, although a human's body is made of clay, he is not simply part of nature; he is made in God's image."[13] White's argument loses sight of the fact that to be created in the image and likeness of God "is to make them be like God in the way He acts, that is in loving kindness, because by that humans will be able to fulfil their responsibility of cultivating and nurturing the newly formed earth and dealing with its nonhuman creatures," as argued by Mensah.[14] Damonte responds to White's argument that "the Christian message held in Genesis, correctly understood, implies that human power over nature is not arbitrary, nor absolute, but it must correspond with the order of creation."[15]

Miranda N. Pillay claims to identify with ecofeminists, yet somehow sees the assignation of "humans over the environment" and "men over women and children" as designed to "operate as hierarchies to oppress and subjugate."[16] However it is also believed that, "ecofeminists reject the 'kingship' model where human beings dominate and subdue the earth.[17]

It is hard to find good reason for the scholarship against Christianity *per se* in ecological matters, while the Tanakh or Hebrew Bible is subscribed to not only by Christianity, but also by the adherents of Judaism and Islam. All three religions claim that the God revealed by the Tanakh is the only God of the earth and sky who created the heavens and the earth and everything in between. Well, we will not make these positions against the assignation of humans the fulcrum of the discussion in this paper but move on to review some of the rules the Creator set in delegating that responsibility for the

---

12. Truelove and Joireman, "Understanding the Relationship," 807.
13. Damonte, "God, the Bible," 32.
14. Mensah, "Uniqueness of Humankind," 34.
15. Damonte, "God, the Bible," 41.
16. Pillay, "Church and the Environment," 185.
17. Pillay, "Church and the Environment," 194.

sustenance of his handy work. We agree with the view that the concept of dominion over every living thing, as implying that humans have the right to dominate and exploit creation, is a misinterpretation of the text of the creation story in Genesis 1. Rather, as argued by the author of "Stewardship as Sharing in God's Life and Mission," "the dominion principle, as the expression of human role in God's creation, can be likened to the relationship a good shepherd has with the sheep."[18] In the rough terrain of the mountains, and with the perils and threats of beasts in the wild, the good shepherd lays down his life for the animals.

Care for the Land

Scattered in the Torah are guidelines for humans to dutifully carry out the responsibility of good stewardship of the earth God handed down to them. I have no intention to go through it all here, but have selected a few which will help make the importance of stewardship well established in a discourse on the environment and care for the land.

a. God first cares for the land

There is a hard truth that we need to acknowledge at this point as we move deeper into the discussion. God has called humans to care for the land he created but it is not that he might take a rest or that God could not do it himself. No! Not at all! For God surely will not ask humans to do anything God will not do himself. God actually will do the hardest bit to maintain the earth God has created. God holds it all together so heavenly bodies and flying avalanches do not crush into each other. We infer from the OT that the hilly country of Palestine in which God settled the Israelites, was not easy to work. It was not as fertile as the Nile Valley which they irrigated and cultivated in the land of slavery. Beyond the Jordan River is a land of mountains and valleys that depend on water from heaven, but the Lord God cared for it; he kept constant watch over it throughout each year. This is the land God gave to Israel. If Israel occupied and remained faithful, the Lord would not curtail his keep, but would send rain and ensure there was grass enough in the fields for the animals and to produce sufficient food for the humans (cf. Deut 11:10-15). God is with humans in the responsibility as stewards and has set rules for its management and use. In the following, we consider a few of the rules and the extent to which inhabitants of the earth have kept them.

---

18. Methodist Church Ghana, "Stewardship as Sharing," 16.

b. The Moral and Ethical Principle

Caring for the land and keeping it right has both moral and ethical considerations. Starting with the birds of the air, for instance, we learn birds do not work the land, they do not sow or reap in the harvest, but they are fed by God (cf. Matt 6:20). God does so from the fields cultivated by humans. Consequently, it is unethical for farmers to go back over their fields after a harvest to glean for the crops that dropped on the land during the harvest. Farmers have a moral obligation not to harvest the crops that grow in their fields by themselves during a Sabbath year because from what grows in the fields and in the wild God feeds the poor amongst them and provides for the wild animals (Lev 25:5-7). This moral and ethical law extends a bit further than just having the poor, the birds, and wild animals in mind. The principle also applies when men go out to war.

"When you lay siege to a city for a long time, fighting against it to capture it, do not destroy its trees by putting an ax to them because you can eat their fruit. Do not cut them down. Are the trees people, that you should besiege them? However, you may cut down trees that you know are not fruit trees and use them to build siege works until the city at war with you falls" (Deut 20:19-20).

Fighting men must not vent their anger on the plantations of the enemy country because they are not at war with trees. Rather, they might find good use for them. If the trees are fruit trees, they will have their food from it and if not so, the trunks of nonfruit trees might be used in temporary construction work in their camps. Over and above those considerations, the wild is a habitation for nonhuman living things (see Gen 1:24-25).

c. The land is a dwelling place

The next principle is that the land is a habitat, not only for humans, but all other living creatures, birds, fishes of the water bodies, and beasts of the wild. It is also a dwelling place for God, who lives amongst his people, and the land must not be polluted. There are prohibitions to prevent humans from polluting the land in order not to inconvenience other inhabitants, including the presence of God living amongst them:

> Do not pollute the land where you are. Bloodshed pollutes the land and atonement cannot be made for the land on which blood has been shed, except by the blood of the one who shed it. Do not defile the land where you live and where I dwell, for I the Lord dwells among the Israelites (Num 35:33-34).

Pollution in the land is forbidden because of its devastative effect, which can hardly be reversed. Whether it is bloodshed, as cited in the text above, or other pollutants such as dirty oil spilt on the land by mechanical operators, polystyrene material (water sachets and wrapper rubber bags), or the indiscriminate application and disposal of chemical substances, liquid waste, or gaseous pollution of the atmosphere by factories.[19] The rule is that "Those who destroy the land will also be destroyed by the Creator" (Rev 11:18).

Humans must not look on while the land turns patchy, and vegetation and green fields wither for lack of maintenance. It is a shame to see institutions of higher learning teaching about protecting the environment, while some of their lawns and playing fields are in such poor shape, so also is it to the agencies for maintaining and keeping green children's parks and street islands. Humans must take advantage of Abhor Days, reforestation programs to check desertification, and embark on good agricultural practices to keep the earth in good shape. Otherwise, for the sheer wickedness of those who refuse to care for the land, "the animals and birds perish" (Jer 12:4). As the saying goes, "the kindest acts of the wicked are cruel but the righteous care for the needs of their animals" (Prov 12:10). The implication of these sayings are that the land, birds, and animals must not be destroyed by any means because humanity does not occupy the land alone. Wildlife, aquatic life, microorganisms, and birds of the air all live in it. God also dwells amongst his people. God pitches his tent among humans "and he dwells with them" (cf. Rev 21:3). We now turn to cultivatable land arising from the mandate to have for food "every seed-bearing plant on the face of the whole earth . . ." (Gen 1:29).

d. Rules for agricultural land use.

In Ghana, one area of human activity in which the land suffers quietly is agriculture. Vast expanses of land are cleared year after year in order to establish farms of all sorts. Mechanical agriculture and valued timber felled for lumber all take their toll. Chainsaw operators continue to deplete the forest in many places, and forest guards hardly go to confront them, especially on nonworking days.

---

19. Nyanor, "EPA Shuts Down Plastic Factory." Nyanor reported on February 13, 2018 that Osudoku Senior High School had closed down due to pollution from the Shinefeel Ghana Company Ltd. The head mistress explained the school could no longer contain the harmful smoke from the factory. It came to light the company did not have a license to operate in plastics.

Many years ago, shifting cultivation was practiced widely in Africa, with its attendant slashing and burning. African countries were criticized by the West for wasting land and destroying the ecosystem. How ironic that such criticism paved the way for Western agricultural industries to export their fertilizers and other chemicals to new markets in Africa! Leaders of African nations did not take into serious consideration the implications of the action then before they embraced the technology. Meanwhile, the Bible had a solution for maintaining the land for the Jews and saving it from depleting its fertility. One of the provisions is in fallowing the land for a period as was in a shifting cultivation system. The principal rule was,

> For six years you are to sow your fields and harvest the crops but during the seventh year let the land lie unplowed and unused. The poor among your people may get food from it, and the wild animals may eat what is left. Do the same with your vineyard and your olive grove (Exod 23:10-11; cf. Lev 25:3-7).

The fallow year took care of the landless people amongst them, the wild animals, and ecosystem. As observed by F. F. Bruce, the "provision guarded against the exhaustion of the soil by too-intensive cultivation."[20] The land got naturally fertilized by the droppings of wild growth and animals having a field day because of the absence of regular human activity. Nocturnal creatures got a whole year to turn the land over without human interruption. Ant hills and termite mounds were built overnight and others such as the grass cutter, the badger, and rats burrowed the earth without any disturbance.

The more fascinating reason enjoining users of the land to fallow it in the seventh year from any activity was that "The land itself must observe Sabbath to the Lord" (Lev 25:2b) and also "have a year of Sabbath rest" (v. 4a). The land owes obedience to God the Creator to observe the demands of his decrees just as humans do. The land requires the cooperation of humans to make it possible. Otherwise sooner or later, the land will sleep on the farmer's effort and fail to produce its yield in abundance. Humans have an ethical responsibility to allow the earth to obey its maker (v. 5). It will be morally wrong to observe Sabbaths every week and yet refuse the earth rest every seventh year. When the land rests, it will recuperate and kick back into productive action again.

The Sabbath rest was to be repeated in the fiftieth year, called the Jubilee, after seven successive Sabbaths (Lev 25:8). The jubilee required all the demands of a seventh-year Sabbath (vv. 6-7; 11, 12), and in addition

---

20. Bruce, "Bible and the Environment," 20.

made provision for the return of landed property to their owners. The celebration commenced with the sounding of the trumpet throughout the land, calling attention to the consecration of the jubilee and the proclamation of liberty that required inhabitants to return to their original properties. Whatever the land produced in the jubilee year could be eaten (v. 7) but had to be taken directly from the fields; owners were not to go out and harvest in order to stock.

e. Sale of Land

Two important reasons are given for the prohibition to sell land: "land must not be sold permanently, because the land is mine and you reside in my land as foreigners and strangers" (Lev 25:23). First, in the jubilee declaration of liberty, it is made clear to all inhabitants, "each of you is to return to your family property and to your own clan" (v. 10b). This rule ensures that individuals are not alienated from the right to land. If a people are dispossessed of that right, they may engage in subsistent farming, meaning they would be deprived and finally lead to poverty because they are landless individuals. F. F. Bruce sees in this rule an evolving "principle that land should not be perpetually alienated from the group that owned it" that subsequently became "firmly entrenched in Israelite law."[21]

The second reason the land must not be sold permanently is that it belongs to God the Creator. The humans are only temporary custodians and inhabitants of the earth who do not have the *bona fide* right to dispose of what belongs to the Creator. If they mortgage it, they must also have plans to redeem it (cf. Lev 25:24). In this matter, no one is to take advantage of another; the rules governing temporary transfer of land and property must be adhered to (cf. vv. 14–17).

In Ghana, chiefs used to give land to prospective settlers and developers; churches were the most frequent beneficiaries of free land. If the land was not developed within a certain time period, the title deed reverted to the original owners. Outright sale or purchase did not apply but now it has become the order of the day so that even churches and institutions have to buy land. Many people in some of the communities do not yet realize that once land has been sold outright, owners cease to have right of entry. This has brought about the engagement of land guards, armed by their employers to maim or kill anyone who comes into a land the owners want to resell at a point in time. The situation becomes chaotic especially when girded by political interests.

---

21. Bruce, "Bible and Environment," 21.

Until the rules of the Creator concerning land transfers are adhered to in some way, peace and tranquility will continue to elude many societies (cf. v. 18). Here we expect theologians to do more education to bring understanding of the matter to their people. Bruce has argued that "those who possessed it received it from him (God) as a heritage, and it was not for them to dispose of it at random"[22] Similarly, in the view of the traditional religionist, "land is nobody's property. It is communally owned and the chief or family-head holds it in trust on behalf of the extended family. No one has the right to sell part of it because it is a legacy from ancestors."[23] Regrettably, "for a pittance chiefs [now] sell ancestral lands to foreigners in the name of so-called African hospitality but once the land is sold there is no way anybody can control how it is used and for what purpose."[24] The OT makes clear that anyone who sells inherited land outright reduces the clan's land size by the measure of the portion sold (see Num 36:3–7). Because we do not have the luxury of time and space to discuss this point further, let the following two examples cited from the Bible suffice as illustrations of the point being made.

The first is the story of the five daughters of Zolophehad (Num 36:1–12). Their father died without a son, and the daughters did not inherit their father's land at the time. Consequently, they were denied any portion of land that belonged to their father. When the matter came up before Moses for consideration, the direction was to give them inheritance from amongst their father's brothers (Num 27:1–11), but they must also marry from within his tribe (Num 36:8–12). This decision ensured that the property of one clan was not carried to another through marriage.

The second is the story of Naboth, who refused to sell his vineyard to King Ahab (1 Kgs 21:3). It reminds us of the moral questioning of one who sells land knowing he or she did not have a perpetual right to the land and an ethical issue for those who buy, knowing that this seller did not have ultimate control over the land. This example too is at the heart of many disputes over land in many communities and needs to be given attention. Indeed, as Damonte rightly put it, "we do not possess nature, but we have received it from past generations and we have to preserve it for future generations and for the necessities of poor people"[25] and the wild animals and creatures who also live in the land as humans do.

22. Bruce, "Bible and Environment," 21.
23. Opoku-Agyemang et al., "Akan Traditional Beliefs," 143.
24. Opoku-Agyemang et al., "Akan Traditional Beliefs," 143.
25. Damonte, "God, the Bible," 41.

Human Activity and Degradation of the Earth and Environment

"Is it not enough for you to feed on the good pasture? Must you also trample the rest of your pasture with your feet? Is it not enough for you to drink clear water? Must you also muddy the rest with your feet? Must my flock feed on what you have trampled and drink what you have muddied with your feet?" (Ezek 34:18–19)

We must point out here that we do not mean all human activity amounts to a degradation of the earth and environment. Indeed, in the creation story, when God commissioned humans as custodians of the earth, God also provided for legitimate actions in which humans have to feed on vegetation (Gen 1:29) and "everything that lives and moves" (9:5). God also gave elaborate rules and principles for maintaining a sustainable earth. We will shortly argue that it is a breach of those rules and principles that constitute abuse of the earth and environment. Presently, we will move on to discuss some of the human activities that appear to be in excess of the provisions and guidelines given to sustain the earth. We will not do much of celestial degradation beyond this point. Water is life on earth; that will be our starting point.

a. Water and air pollution

The two spheres of creation—terrestrial and celestial—have been affected by human action. For instance, there is a gaping hole in the ozone layer. In many countries, pollution of the atmosphere has been so high that it is no longer safe to drink rain water. In a panel discussion on climate change and its impact on public health at the University of Cape Coast, Gina McCarthy, a US Environmental Protection Agency (EPA) administrator noted that "17,000 die annually from air pollution in Ghana as a result of exposure to air pollution" and "children and women are the most affected victims." She also indicated "these deaths are the result of high amounts of hydrochloric acid being released into the atmosphere."[26] Similarly, sulfur dioxide emitted from the combustion of fossil fuels like coal, petroleum, and other factory combustibles are a major cause of air pollution. In Ghana, we cannot say that rain water is as safe as it used to be.

---

26. Crawford and Knight, "Ghana Sustainable Fisheries Management Project" para. 3. https://www.crc.uri.edu/stories_page/us-epa-administrator-gina-mccarthy-visits-usaidghana-sfmp/.

The water situation is rather precarious and serious, as the Social Affairs Committee of the Canadian Catholic Bishops Conference points out, "without water everything dies" because "water is the basic element through which all life forms emerge, exist and flourish. Water is the lifeblood of the planet."[27] How then are churches and theologians in the country usually silent and why do they not raise their voices about what is happening to society? Theologians and Christian men and women usually claim to have new life because they were born again when they believed in Christ and emerged out of water through baptism (cf. Acts 2:36–38). They have continued to exist and to flourish because they depend on water daily. How can such people remain unconcerned while water appears to be the most abused element in rural society? In this country, water suffers from three major human activities, solid and liquid waste disposal, chemical use in agriculture, and galamsey.

Solid and liquid waste disposal

A few years back many people went to the stream to fetch water daily and it is still the case in some communities in Ghana. However, today it is not very safe because several water bodies have been contaminated and polluted by human actions. The Korle Lagoon in the city of Accra has fish but not many admire the idea of having to buy fish from the market coming from that source. The Densu River used to be a great source of fresh water for many villages along its tributaries and main stream flowing all the way, till its waters were treated and ended up in mansions of Accra through the taps. The water company treats this water when it gets to the Weija Dam in Accra, but before it gets there a lot of it has been consumed along its course by rural communities in a contaminated state. No wonder there is a perennial occurrence of cholera along the Densu basin and in parts of Accra.

Many people in the cities of Ghana have no idea about the garbage and sluice tipped in the vicinity river courses and how its waste waters seep into the basin before it reaches the dam where the company turns it into tap water. Drastic measures need to be taken to control the mess in the Densu River, which takes its source from a walking distance of a hostel of the All Nations University College at the Main Campus in Koforidua.

The widespread pollution of water bodies has made safe drinking water very expensive. In this regard, an observation made by the Canada Catholic Bishops Council shows relevance: "some persons living in urban slums in poor countries are forced to pay between four and one hundred times more

27. Social Affairs Commission, "You Love All That Exists," 151.

for water than their middle and upper class fellow citizens."[28] People who can afford it do not buy all the water they need in their homes; they have a mechanized borehole right in there, well filtered with the latest technology. The poor people go out to buy bottled or sachet water in the street.

Agriculture

There is more trouble when we come to agriculture. People who cannot read the small writings on an agro-chemical product are farming along streams. Meanwhile, people down the stream in the next village fetch all their water from such polluted streams. What a tragedy that technical agricultural extension services which thrived in Ghana seem no longer to be so at a time when it is needed most. Indiscriminate application of agricultural chemicals and fertilizers are on the rise to meet the demands of the market. We must ask the question: How does the rampant and indiscriminate application of agro-chemicals affect subterranean water in Ghana?

Similarly, a hydrologist with the Centre for Scientific and Industrial Research of the Water Research Institute, Dr. Anthony A. Duah, speaking on the "Effects of Galamsey on the Environment" made the point that Ghana's available water resources were enough compared to its current population size in the 1990s. However, in view of the present destruction caused by galamsey, the nation's water resources are "threatened by activities of illegal miners. Improper solid and liquid water disposal, poor land use practices and sand wining, as well as, deforestation activities."[29]

There is a World Bank Project designed to mitigate some of the problems directly or indirectly associated with water and the land. For instance, there is the Sustainable Land and Water Management project, the objective of which is to expand the area under sustainable land and water management practices in selected watersheds. It is envisaged that the project will promote good practices to reduce land degradation and enhance maintenance of biodiversity in the Kulpawn-Sissili and Red Volta watersheds.[30] While this is a great project, one wonders how far it will go in order to catch up with the spate of degradation.

Likewise, Care International, which also began work in Ghana in 1994, develops projects on livelihood activities with communities and marginalized groups to help them build and manage the natural base over the long term, particularly in northern Ghana where "Food security and nonsustainable

28. Social Affairs Commission, "You Love All That Exists," 151.
29. Ghana News Agency, "Gov't to Spend $100m," para. 8.
30. http://projects.worldbank.org/P132100?lang=en.

land use are major challenges to the livelihood of people."³¹ These are indeed very good initiatives and it is good to know that the projects are ongoing. However, action by local people and civil society need to be stepped up in order to make the interventions part of the daily activities of the people rather than it being a project isolated from the inputs of the communities. This approach would be most helpful in communities where dreadful degradation is claiming lives due to the insurgent activities of illegal mining operators. For so long, this illegal mining activity has evaded attention and control, until it became a health hazard and a political challenge.

Galamsey

We assume that readers are familiar with the Ghanaian term "galamsey" as the illegal prospecting for minerals, particularly gold and diamonds, as a means of livelihood for some people. It is important to note that galamsey impacts negatively on the land and environment, with very serious implications for the communities where such mining activity takes place. As rightly observed by the Bole District Chief Executive, Veronica Alele Helming, "galamsey is a canker and its effects are detrimental"³² to the environment and affects government expenditure excessively. In 2017, the government initiated stringent measures, including a ban on all small-scale mining to curb the menace of galamsey, and the lifting of the ban in 2018 has brought in concrete measures. At least as a start, the government was "to spend about 100 million dollars to reclaim lands destroyed by activities of illegal miners in the country, through the Multi-sectoral Mining Integrated Project (MMIP)."³³

According to Mawutodzi Kodzo Abissath, "all governments since the 1992 Republican Constitution have made efforts to fight galamsey" but without significant success. Not even the attempt made by President John Dramani Mahama in May 2013, when he "inaugurated a high-powered inter-ministerial committee (the anti galamsey task force)" could "find a permanent solution to galamsey." "The task force could not live up to expectations" because it lacked top-level political will to combat the menace.³⁴

31. Ghana Care Canada—http://care.ca/country/ghana, para. 11.

32. Ghana News Agency, "Gov't to Spend $100m," para. 10. Madam Veronica Alele Helming, the Bole District Chief Executive, made this contribution at the 6th Town Hall Forum organized by the Media Coalition Against Galamsey and the National Commission for Civic Education (NCCE) at Bole.

33. Ghana News Agency, "Gov't to Spend $100m," para. 11.

34. Abissath, "Galamsey Land Reclamation Project," para. 4.

From February 2017, the bold attempt to mitigate galamsey activity was initiated by the Nana Addo Dankwa Akuffo Addo-led government by imposing a ban and sending a task force made up of soldiers and police to arrest all who flouted the ban. The initiative inspired hope and confidence because of the bold and tough decisions taken by the government to deal with the problem, as well as the assertiveness of the then-minister for the sector, Ministry of Land and Natural Resources, John Peter Amenu, to carry out the decisions of government. It was believed that the uncompromising stance of the ministry, if sustained, would take the nation out of the doldrums and quagmire of galamsey. Operators were ordered to pull out from their work sites and to remove all their equipment including bulldozers within a certain time frame and they complied. The exercise generated public and media interest. Subsequently, the media pooled together and took a determined position against galamsey. As indicated by Abissath, on April 4, 2017, the Media Coalition Against Galamsey was launched to sustain the campaign. In spite of the shouting back and forth, and complaints that galamsey workers were being deprived of their livelihood, the government had a strong backing from the public and media front.

On July 10, 2017, at a meeting held by the President and Ministry with traditional leaders, the chiefs, and elders at Kyebi, the matter was drummed home even harder. The chiefs got the message about the measures, and there was no turning back. On July 31, 2017, Operation Vanguard, comprising 400 service personnel, was launched "to terminate galamsey once for all."[35] Notwithstanding the brutal murder of an army officer amongst a team posted to Denkyira Oboasi, surveillance efforts continued.[36]

It is common knowledge that wherever galamsey activity has taken place operators have left in their trail massive destruction to the Ghanaian landscape. The minister of Lands and Natural Resources disclosed in a speech on November 13, 2017 that about 4 percent of the country's land size (which works up to 238,000 sq. miles) has been destroyed by galamsey activity alone.[37] Anthony A. Duah also claimed that among all other activities "destruction of the environment caused by mining activities had reached alarming proportions and needed to be controlled to avoid the negative impact of inadequate water supply, poor water quality, loss of aquatic life

---

35. Abissath, "Galamsey Land Reclamation Project," para. 9

36. The military personnel who was gruesomely murdered at Denkyira Oboasi in the Central Region was Major Maxwell Adams Mahama. Investigations are ongoing, and several arrests have been made (see https://www.todaygh.com/ - May 30, 2017; https://www.graphic.com.gh - Sept. 20, 2017; https://www.graphic.com.gh - Jan. 11, 2018; https://mynewsgh.com/ - May 29, 2018).

37. Abissath, "Galamsey Land Reclamation Project," para. 9

and loss of recreation."[38] So far, the good news is that "the transformation of many muddy and polluted rivers that could not be treated by the Ghana Water Company" as well as "farmlands and forests that were devastated, are gradually being reclaimed and replanted with trees."[39]

The Way Forward

We do not intend to make a list of a series of actions to be taken in order to halt degradation of the land. Such a list would not be exhaustive enough. Rather, our way forward is about who should ensure that the land is kept safe and the integrity of the environment upheld. There are three agencies (and their subsidiaries) to whom this responsibility falls: the government, the religious bodies (the church), and traditional authorities.

The political will of the government and traditional authorities are needed to enforce laws and regulations governing land acquisition and use. Traditional authorities "took definite steps to achieve this aim but their knowledge was always shrouded in mystery and secrecy since its modus operandi is spiritual. It behooves on the government to collaborate with traditional society to 'acquaint itself with the realities behind the mysteries"[40] in order to make scientific sense of the messages communicated through that medium for the proper education of modern society on land issues and discreet administration of lands at that level.

The religious bodies constitute the third estate in the care of the environment who should ensure good stewardship of the land as an assignation from the Creator. As a body politic the three monotheistic religions have responsibility to ensure justice for, and integrity of, all creation. As noted by Opoku-Agyemang et al., "Traditional religion has been conscious of environmental protection since time immemorial" knowing "of the absolute need to preserve the environment for the survival of the human species."[41] The Christian church, for instance, can do two things: advocacy work through its agents and adept teaching of the word of God to persuade and to guide society on how to relate to the world of nature, especially the land. It must be pointed out that there is light at the end of the tunnel as some churches in their discipleship and devotional manuals for its members devote sections to instruction on the environment.[42]

38. Ghana News Agency, "Gov't to Spend $100m," para. 13.
39. Abissath, "Galamsey Land Reclamation Project," para. 10.
40. Opoku-Agemang et al., "Akan Traditional Society," 141.
41. Opoku-Agemang et al, "Akan Traditional Society," 141.
42. See Methodist Church Ghana, "Stewardship as Sharing," 16.

That notwithstanding, a lot more advocacy work could be done through its agencies such as West African Association of Theological Institutions (WAATI) and the Ghana Association of Biblical Exegetes (GABES), The Catholic Secretariat, Ghana Pentecostal Council, Christian Council of Ghana, and the Bible Society of Ghana, to name a few. It is gratifying to note that in 2012, GABES, in its *Journal of African Biblical Studies,* dedicated the 4th volume of the publication to ecology, and West Africa Association of Theological Institutions (WAATI) also helped put this volume together.

WAATI and GABES may not be the official mouthpiece of the church, but they both draw membership of academicians from the churches and are mentors of young theologians leaving their institutions to serve society. What these mentees teach or preach in their ministry reflects the influence of their mentors on them. This is a huge advantage for the promotion of a discourse of the environment. In fact, WAATI and other similar bodies ought to be for modern society what the Masoretic scholars were to the Jews. As "traditionalists they helped to preserve the text of the Hebrew Bible between 600-1000 CE."[43] They wrote their observations in the margins of every page called the marginal masora or at the end of the Bible as final masora.[44] The Masoretes did not create the book, they only kept the tradition of interpretation and advanced the textual tradition of the book for posterity. WAATI would do well as a club of theologians to engage the church and society in discourse whenever a burning issue arises, and provide the needed scholarship and leadership direction. It must not be left solely to media practitioners to give their opinion. By so doing, WAATI will be showing relevance as a professional body, or better, interpreters called to a ministry of enlightening society in the word of God.

Conclusion

As one of the three monotheistic religions, Christianity has enough resources to teach and to direct society on the critical environmental issues that arise from time to time. On the subject of caring for the land, there is such a rich store of information to be gleaned from the Bible for the instruction of society that we can certainly do better than we are doing now. For indeed as the pastoral letter on Christian ecological imperative has it,

> To enter into ever-deeper relationship with God, entails striving to develop right relations with nature and with other human

---

43. Seow, *Grammar for Biblical Hebrew,* 170.
44. Würthwein, *Text of the Old Testament,* 28.

beings. Anyone who accepts the truth about God and creation will also agree that caring for the land is not an option. In the Christian perspective, caring for the land forms an integral part of personal life and life in Society. Not to care for the environment is to ignore the Creator's plan for all creation and results in alienation of the human person.[45]

Society will change its attitude not because we asked them to do so, but because we opened up the Bible and pointed out, "thus says the Lord" to convince and to persuade people. When people become aware of the love God has for his creation, they might rethink how to handle the created world that God put in their charge.

A principal teaching of the Bible to which humanity's attention ought to be drawn to is the insight of the apostle concerning the renewal and redemption of all creation. All adherents of the three monotheistic religions—Judaism, Christianity and Islam—look forward to a kingdom of heaven coming their way. But not much emphasis has been made concerning the apostolic insight that creation also awaits redemption in order to be the new heaven and the new earth (Rev 21:1)

The sin of humanity had disastrous consequences. The female will experience severe pains in child-bearing (cf. Gen 3:16), and the man, for listening to the deception carried by his wife, will feed his family through painful toil and the sweat of his brow all the days of his life (vv. 17–19). For their sake, the land too shares in the punishment: "Cursed is the ground because of you" (v. 17b). Thus, from the beginning, both humans and the land suffered curses, and both require redemption. It is against this backdrop that "Paul's comments in Rom 8:19–24 on the cosmic aspects of the redemption that Christ has procured for his people"[46] make sense:

> For the creation waits in eager expectation for the children of God to be revealed. For the creation was subjected to frustration, not by its own choice, but by the will of the one who subjected it, in hope that the creation itself will be liberated from its bondage to decay and brought into freedom and glory of the children of God. We know that the whole creation has been groaning as in the pains of childbirth right up to the present time. Not only so, but we ourselves, who have the first fruits of the Spirit groan inwardly as we wait eagerly for our adoption to sonship, the redemption of our bodies. For in this hope we were saved (Rom 8:19–24a; NIV).

---

45. Social Affairs Commission, "You Love All That Exists," 147.
46. Bruce, "Bible and Environment," 29.

On that day of the great resurrection, it is not only humans who will experience liberation and fulfillment, for the creation too is waiting with eager longing to be "liberated from its bondage to decay and brought into the freedom and glory of the children of God" (cf. v. 22). The following comment by F. F. Bruce is instructive: "Like man, creation must be redeemed—because, like man, creation has undergone a fall."[47] The land is trailing behind the children of God, having suffered the curse with them in the garden of Eden. When the Lamb of God appears, there will no longer be any curse (cf. Rev 22:3a). The moral obligation is that the land requires the attention of all humans that it might be made ready for the day of redemption (Rev 22:1–6).

Bibliography

Abissath, Mawutodzi Kodzo. "Galamsey Land Reclamation Project: Don't Ignore the Ghanaian Media." http://ghana.gov.gh/index.php/media-center/features/4209-galamsey-land-reclamatio-project-don-t-ignore-the-ghanaian-media.

Bruce F. F. "The Bible and the Environment." In *Living and Active Word of God: Essays in Honour of Samuel Schultz*, edited by Morris Inc & Ronald Youngblood, 15–29. Winona Lake, IN: Eisenbrauns, 1983.

Chang, Kenneth. "NASA's InSight Mission has Touched Down on Mars to Study the Red Planet's Deep Secrets." https://www.nytimes.com/2018/11/26/science/nasa-insight-mars-landing.html?searchResultPosition=1.

Crawford, Brian, and Knight, Maurice. "Ghana Sustainable Fisheries Management Project." https://www.crc.uri.edu/stories_page/us-epa-administrator-gina-mccarthy-visits-usaidghana-sfmp/.

Damonte, Marco. "God, the Bible and the Environment: An Historical Excursus on the Relationship between Christian Religion and Ecology." *Relations* 5.1 (2019): 27–43.

Ghana News Agency. "Gov't to Spend $100m on Land Reclamation." http://citifmonline.com/2017/12/govt-spend-100m-land-reclamation/.

Harris, Stephen L. *Understanding the Bible*. 6th ed. Boston: McGraw Hill, 2003.

Hinsdale, M. A. "Heeding the Voices." In *The Embrace of God: Feminist Approaches to Theological Anthropology*, edited by Graff A. O'Hara, 22–50. Maryknoll, NY: Orbis, 1995.

Longman, Tremper, III, and Raymond B. Dillard. *An Introduction to the Old Testament*. 2nd ed. Grand Rapids: Zondervan, 2006.

Mensah, Augustine M. "The Uniqueness of Humankind: A Close Reading of Genesis Creation Accounts." *Journal of African Biblical Studies* 4 (2012) 25–34

Methodist Church Ghana. "Stewardship as Sharing in God's Life and Mission." *Weekly Bible Lessons* (Jan-June 2018) 14-18.

Millay, Miranda. "The Church and the Environment: On Being down to Earth in a Consumerist Era." *Scriptura* 107 (2011) 184–98.

---

47. Bruce, "Bible and Environment," 29.

Nyanor, George. "EPA Shuts Down Plastic Factory." www.myjoyonlline.com/news/2018/february-13th/epa-orders-closure-of-platic-factory-causing-pllotion-at-at-asutuare.php.

Seow, C. L. *A Grammar for Biblical Hebrew*. Rev. ed. Cincinnati: Compublishing, 1995.

Social Affairs Commission, Canadian Conference of Catholic Bishops. "You Have All That Exists . . . All Things are Yours, God of Life." *Mission* 11 (2004) 147–56.

Truelove, Heather Barnes, and Jeff Joireman. "Understanding the Relationship between Christian Orthodoxy and Environmentalism: The Mediating Role of Perceived Environmental Consequences." *Environment and Behavior* 41.6 (2009) 806–20.

White, Lynn, Jr. "The Historical Roots of Our Ecological Crisis." *Science* 155 (1967) 1203–7.

The World Bank, "Sustainable Land and Water Management." http://projects.worldbank.org/P132100?lang=en.

Würthwein, Ernst. *The Test of The Old Testament*. 2nd ed. Translated by Erroll F. Rhodes. Grand Rapids: Eerdmans, 1995.

Part I

Toward an Agenda for African Ecotheology in African Theological Studies

MARK S. AIDOO

Trinity Theological Seminary, Legon

Abstract

For theological study in Africa to influence the needs of African people, it needs to inculcate an integrated ecotheological agenda from an African perspective so that it can equip theological students to be partners in caring for land, waterbodies, and the environment. African nations are struggling to find answers to ecological issues. This paper examines the complex intersections between the biblical mandate of nature care in Genesis 2:4b–17 and African identity to illustrate how a distinct patterning of theological studies can emphasize human vocation and care for the land. Three benefits of African ecotheology are discussed. First, it will enhance the African identity located in relationship with nature from an African perspective. Second, it will explicate on the mandate of God for humanity not only to save souls but to promote life in abundance. Third, it will enhance African spirituality. The paper then advocates that theological students need to equip themselves to stand by African governments in their effort to regulate environmental risks associated with land care including pollution of our oceans, water, and wildlife.

Introduction

THIS PAPER PROPOSES THAT theological studies in Africa are incomplete without participation in the public discourse of land and environmental issues and an integrated study of African ecotheology. It argues that theological and religious studies in Africa must promote an integrated ecotheological agenda in its missional outlook that equips students of the Bible to be partners in caring for land, waterbodies, and the environment. By examining the complex intersections between the biblical mandate of nature care in Genesis 2:4b–17 and African identity and spirituality, the paper illustrates how African ecotheology becomes primary to the Christian salvation message. It argues that the call on humanity to till and care for the land as a sacred mandate ought to take precedence in Christian ministry.

The Nature of Ecotheology

It is crucial to point out that the call for ecotheology is not the same as a call for environmental ethics. Although environmental ethics is multidisciplinary, reflecting on attitudes, knowledge, and philosophical considerations on welfare and values, its focal point is on what human ought to do.[1] In the words of Holmes Rolston, "Environmental ethics has to be directed to human dominated, managed, disturbed (and often degraded) landscapes. Such a land ethic must be informed about ecosystem health, but more focused on human ecology, on political ecology."[2] In other words, it is the ethical issues related to human actions about the environment that become the focus. Issues about nature are relegated to the background instead of playing an equally significant part in discussions. Rolston adds that "Environmental ethics cannot be an ecosystem ethic pure and simple; there is only an ethic about humans relating to their ecosystems, in the economies in which they live.[3] Whereas environmental ethics focuses primarily on human beings, the central feature of ecotheology is not only about human beings. It is about God and all creation. A study about ecotheology should not be dependent on one's personal convictions, beliefs, and philosophical suppositions. For African Christians, such a study should be dependent on the Bible with an all-inclusive concern for the well-being of all creation.

Generally, ecotheology has been a subject under systematic theology. However, Paul Santmire thinks otherwise. He argues "Christian theology

---

1. Light, "Urban Ecological Citizenship," 44-63.
2. Rolston, "Environmental Ethics," 524.
3. Rolston, "Environmental Ethics," 524.

never has had, nor should it have, a substantive ecological dimension. These writers are convinced that Christian theology must focus primarily—even exclusively—on human history, not on the history of nature. A substantive Christian theology of nature, in their view, is a contradiction in terms."[4] I partly agree with Santmire that it needs to be noted that human history has very little to do with land and nature. The root word for "eco" being *oikos* ("household") encompasses the habitat of humanity which is the world of nature. Everything that belongs to the household should be part of ecotheological studies. The concept of the land embodies the space for humanity, waterbodies, animals, trees, and the earth as the footstool of God. In the Old Testament, land is used as a synonym for earth. Walter Brueggemann argues that land is not a space that echoes emptiness, waiting to be filled, but a place where history is concrete. He says:

> The land for which Israel yearns and which it remembers is never unclaimed space but is always a place with Yahweh, a place well filled with memories of life with him and promise from him and vows to him. It is land that provides the central assurance to Israel of the historicity, that it will be and always must be concerned with actual rootage in a place which is a repository for commitment and identity.[5]

Since the late twentieth century, other theologians (not necessarily systematicians) have begun to pay attention to ecotheology that highlights the totality of God's creation. In my view, ecotheology has become a discipline that combines the disciplines of ecology, theology, and life, and so it must equally be examined through the lenses of biblical theology, practical theology, history, and African thought. Such a study will highlight the interrelationships between religion, sociology, and nature in the light of environmental concerns. It is not simply about humanities. In this light, ecotheological studies shape our understanding of God, the place of human beings, and creation in God's plan, as well as a commitment to social justice. Such a relationship is a matter of concern to every task of theological studies. It is in this light that the proposal that every discipline in religious and theological studies must include aspects of ecotheology in its curriculum come to the fore. Ecotheology should not simply be a core or an elective course. Such a proposal is not simply to champion a subject on ecotheology in the educational structure, but a pragmatic agenda whereby ecotheological issues become distinct features in the syllabus of every theological discipline. I agree with Ernst Conradie when he says:

4. Santmire, *Travail of Nature*, 3.
5. Brueggemann, *Land:* 5-6.

> On this basis one may argue that ecotheology is not so much one form of doing theology alongside other forms of self-consciously contextual theology. It has become a dimension of all theological reflection in the sense that an environmental awareness may be relevant to almost every conceivable topic raised. The same may apply to feminist and liberation theologies: all theologies should be gender-sensitive and liberatory.[6]

That is to say, one cannot study the Old Testament or New Testament without reflecting on ecotheology. The imaginative silence of the New Testament on land care does not mean biblical theologians must do the same. Likewise, it will not be enough to study practical theology like pastoral care, counseling, Christian education, and preaching without helping the student to know how to address issues of ecology. Mission studies and evangelism should engage students on ecotheological issues. One should not study *missio dei* only in terms of human salvation. Jürgen Moltmann understands mission as a concern for all nature when he says:

> The promise of the kingdom of God in which all things attain to right, to life, to freedom, and to truth, is not exclusive but inclusive. And so, too, its love, its neighbourliness and its sympathy are inclusive, excluding nothing, but embracing in hope everything wherein God will be all in all. The *pro-missio* of the kingdom is the ground of the *missio* of love to the world.[7]

Similarly, church history must trace the history of missionaries and churches concerning their stance on ecotheology, and how they included land care in their vision and mission. As Clive Pearson writes, "The dilemma for an ecotheology is that very little in-depth historical research has been done to date on how actual Christian communities have affected or responded to their own environments. There are not the ecological equivalents of postcolonial histories of Christian missions to nonwestern societies."[8]

The challenge will rightly fall on the capacity of theologians and faculty members who must inculcate ecotheology in their lectures. Some decades ago, merely having a first degree qualified one to teach in a theological institution. Today, one needs a research masters, postgraduate, or doctoral degree in order to teach. The basic requirement for appointment has been so refined to make those with vocational degrees like Master of Divinity and Doctor of Divinity or Doctor of Ministry not qualified to train students in

---

6. Conradie, "Contemporary Challenges to Christian Ecotheology," 108.
7. Moltmann, *Theology of Hope*, 224.
8. Pearson, "Electing to Do Ecotheology," 19.

theological education. Experience is no longer the best teacher. A serious academic inquiry on ecotheological issues from an African perspective will go a long way toward helping students to know who they are and further position them to participate in God's business of caring for the earth.

Theological Education as Life Concerns

There seems to be a growing trend that some theological students and church leaders really do not care about the impact their studies could help them to make. They badly need "PhD" next to their name for prestige and not to make a contribution to the field through writing and teaching.[9] When such people are deemed qualified to train pastors for churches, what impact will they make on the students? What contribution can they make to the church beyond the limits of their job as teachers? Are theological institutions in Africa seen as a mile wide but one inch deep?

Theological education certainly places a high value on a holistic approach to training for the ministry. Such a view combines the elements of cognition, volition, and affection. There is no doubt that the emphasis on the heart over the head and the misplacement of hand within the structure of theological education poses a challenge to the church. Theological education should not give less prominence to praxis than reflection and meditation either. To be a practical Christian, there is the need to engage in ways that promote care for nature, one's surroundings, and one's own body.

Thomas Oduro mentions seven types of theological education. These are: (1) formal or Western institutional type; (2) periodic type–seminars and workshop; (3) mentorship/apprenticeship/discipleship; (4) theological education by extension; (5) distant learning; (6) self-taught one; and (7) regular Christian education in the church.[10] This means that there are various opportunities and platforms upon which theological education can be possible. Oduro notes that among the AICs in Ghana, the popular type of theological education and training is the mentorship/apprenticeship/discipleship one, yet the mentor does not adequately prepare the mentees to face the challenges of modern-day issues in Ghana.[11] The same observation can be made about the mainline churches where formal institutional training is interspersed with practical training under senior ministers. In the Methodist tradition, a ministerial candidate or probationer is expected to study the denominational traditions and ministerial practice in the church

---

9. See also Crane, "Equipping the Transient for Ministry," 3.
10. Oduro, "Theological Education and Training," 4–5.
11. Oduro, "Theological Education and Training," 11.

setting. Such a training is usually between five and six years. Opportunity is provided for superintendent ministers to mentor and train candidates and probationers under their care in the ministry. One wonders if these superintendent ministers are up to the task and can offer useful guidance in terms of ecological issues. Again, most of the candidates who study under the theological education by extension hardly ever experience conscious mentoring from the superintendent ministers. They are hardly ever invited for a one-to-one training session. How then can theological institutions fill the gap of providing the necessary tools for mentors in the fields so that they could be effective mentors?

In education, scholars like David Krathwohl, Benjamin Bloom, and Bertram Masia have advocated for the taxonomy of the affective domain in education based on the concept of internalization, and argue that the more a value or attitude is internalized, the more it affects behavior.[12] The affective taxonomy attempts to map noncognitive learning in a parallel fashion to Bloom's taxonomy of cognitive learning. Affective taxonomy uses internalization, which ranges from "receiving" and "responding" to affective phenomena of "valuing" that involves personal acceptance of values. At higher levels of education, internalizing values or attitudes gives the ability to create value systems, prioritize values, and resolve conflicts, resulting in consistent and predictable behavior characteristic of the individual. Undoubtedly, theological education is supposed to follow the design of affective taxonomy that helps students to internalize certain values. Unfortunately, how much of the cherished African values and traditions become internalized through theological studies in Africa for students to exhibit predictable behaviors within the church and state is a question to be answered.

Charles M. Reigeluth and Barbara L. Martin have built on the affective domain by developing a six-fold taxonomy for understanding learning in the affective domain, with an in-depth discussion on "knowledge," "skills," and "attitudes."[13] Their model, which includes emotional, moral, social, spiritual, aesthetic, and motivational development provides a place to begin to think about how to enhance affective development within the context of theological education. These six-fold dimensions of learning, taken together with cognitive learning and skills, give an integrated approach of formation where spiritual formation, intellectual knowledge, and emotional maturity are developed through experience and relationship. Significantly, these dimensions of learning focus on the individual development to the neglect of social demands put forward by church institutions. That is to say, what

---

12. Krathwohl et al., *Taxonomy of Education Objectives Handbook II*.
13. Martin and Reigeluth, "Affective Education," 493.

the church must do in society is not an agenda when it comes to individual learning skills. Hence, the individual student must find ways to apply the knowledge, cast a vision, and explore ways to affect society through the church. Perhaps the advocacy of Karl Marx is worth noting: "Men must be in a position to live in order to be able to 'make history.'"[14] People who make history affect their world with a positive influence. Such a view is in contrast to the Hegelian ideology that looks at world history "in the realm of pure thought" within the sphere of "conceptions, thoughts, ideas, in fact all the products of consciousness."[15] Marx is right in advocating for "real individuals, their activity and the material conditions under which they live."[16] Attention should therefore be placed on how one's personal affective development will really affect others and transform society and nature.

The relationship between church and state has always been a checkered one. Post-colonial governments in Africa found it difficult to see the relevance of theological seminaries in a social world. Institutions of higher learning have restricted religion and theology to the humanities in secular universities. This does not mean religion and theology should restrict itself to issues about human beings to the detriment of society and nature. It is noteworthy to observe that today, some African governments are prepared to accredit seminaries. It is now required by law in countries like Ghana, Nigeria, South Africa, and Kenya for all seminaries to be registered with the government and subjected to a quality assurance audit in order to be accredited. It may not be out of place to say that theological institutions in Africa have not impacted society as they should. Society finds reason to set the agenda for theological institutions to join the fray and to approve its curricula. That should give theological institutions the urge to play a more positive role in social issues as well as the care for nature. As Terrence Fretheim observes, "It is ironic that the impetus for the church's concern for matters creational has come largely from secular sources. That in itself is a considerable witness to the importance of creation theology: God the Creator is pervasively at work in the larger culture, often independent of the church, and that divine activity has had good effects."[17] As long as the church waits for society to provide the agenda for social action, it cannot play its role as the salt and light of the earth (Matt 5:13–16; Mark 9:49–50). Dietrich Werner makes a critical observation about the lack of attention to

---

14. Marx, *German Ideology,* 48.
15. Marx, *German Ideology,* 39–41.
16. Marx, *German Ideology,* 42.
17. Fretheim, "Preaching Creation," 76.

the study of ecotheology in theological institutions although much studies have been done on ecotheology. He says:

> Although much research and project related work has been done already on these topics in some regions in World Christianity, only in a few institutions of theological education and Christian leadership development that issues of eco-theology, climate justice and food security form part of the regular curriculum of instruction and training or inform theological formation processes in an integrated perspective, especially in and in exchange with the global South. A great potential therefore lies in discovering existing resources for the transfer of knowledge and wisdom from all contexts/regions.[18]

The imaginary walls that separate the church and their own theological institution cannot be overemphasized. The voices of theologians and the prophetic voice of the church in national issues hardly become distinct. The theologians and institutions hardly influence the churches that set them up with theological information. In Ghana, when the government heeded the advocacy of the media to ban the activities of galamsey in February 2017,[19] it took a while before the churches started releasing communiques on the issue in support of the government. The church since then is yet to set its own agenda in the fight against destruction of the land and waterbodies. No wonder Dirk Smit writes: "Theological study becomes an individualistic enterprise, done for private reasons, and not because of a sense of calling, a need to serve, a willingness to be formed, or a feeling of accountability. The theological study often fulfils private needs and estranges the students from the church communities and from society or public life at large."[20] There is the need, therefore, to design theological education in Africa to prepare students to play an active role in African religious, cultural, social, and public life. Dirk J. Smit further makes the point succinctly when he avers:

> It becomes almost impossible to fulfil the overarching, integrating role that religion played in earlier societies. Society becomes secularized. This does not mean that there is no place for religion. On the contrary, religion can still be extremely valuable, important, and popular, but then it must restrict itself to the private sphere of the individual's personal, often intimate life. In a modern, secular society religion is privatized. It loses its place in public life—often voluntarily and gladly. Its connections to

---

18. Werner, "Introduction," 13, 14.
19. This is local terminology referring to operations of illegal small-scale mining.
20. Smit, "Modernity and Theological Education," 86.

other subsystems, such as politics, economic life, and public media, the legal system, public education, etcetera, are seriously threatened, and often made impossible by walls of separation.[21]

Theological institutions need extensive networks with dioceses, circuits, and societies, as well as connections with their leadership of the church, so they can be of relevance beyond the classroom. The professors and lecturers need to design programs for their local communities and provide more informal pastoral training and theological education.[22] Informal theological education that goes to the churches, pastors, and congregation where they are, rather than asking them to come to it, can be an ideal space where the real training for the church and society will take place. The call is on theological educators in Africa to see how to prepare students to go back to the old ways where religious functionaries James Kwegyir Aggrey and Gaddiel Acquaah played active roles in national and public life.

With the advent of Pentecostal and charismatic ministries, newer forms of Bible training schools, which graduate students after three to six months of study, keep springing up. Their emphasis is more on spirituality and introduction to the Bible. Most members of the Pentecostal and charismatic churches receive the message of the Bible from pastors and preachers who have no theological education. That is why there must be a constant call for relevant theological studies that addresses the modern challenges of African churches. The alienation between academia and practice, knowledge and ordinary life, continues to be a bane in Africa. The challenge in adhering to the knowledge gained in the Bible to reflect African life issues cannot be postponed to any distant future. One agrees with Teresa Okure that African theologians must not, as a cultural rule, start with the issue of methodology in their theological reflections because doing theology is not just a matter of a correct method, but rather life and life concerns of oneself and the peoples.[23]

The Issues at Stake in Ghana

A study of ecotheology in Ghanaian seminaries and theological institutions must engage issues that confront Ghanaian people and economy. It is believed that presently in the Accra Metropolitan area, for instance, about 2,500 tons of waste is generated daily, but less than half of it is collected

21. Smit, "Modernity and Theological Education," 84.

22. Crane, "Equipping the Transient for Ministry," 9. See also Sills, *Hearts, Heads, and Hands*.

23. Okure, "Feminist Interpretations in Africa," 77.

and sent to landfill sites. Unfortunately, all the landfill sites of the big cities in Ghana are full or filling up, creating another dilemma of where to keep the waste. In the less developed districts and villages, waste is burnt, compounding air pollution and climate change. The rest of the waste which is not collected remains filth in the communities, leading to health risks, choked gutters, and their attendant problems. In fact, the 2014 cholera outbreak in Ghana, which recorded over 17,000 cases with 150 deaths, was caused by filth. In January 2013, the Ghana Health Service declared a cholera outbreak in the Ashanti Region, and as a result eighteen deaths and 310 cases were registered in the region.

Air pollution seems to be nobody's business in Ghana. Foul smells from filth and piles of garbage that fills the air from landfill sites and unattended waste in the communities is a common thing in Ghana. In some places, liquid waste seeps into homes, schools, and public places. Choked gutters become fertile grounds for mosquito breeding and floods. But very little has been said or done to curtail these problems.

It is common knowledge that Ghana is struggling to find answers to the issue of polythene waste. Modernity and propaganda have made Ghanaians shun the valuable medicinal leaves hitherto used to wrap food. Currently, most food and water packages are polyethylene products which form about a quarter of the waste generated in the country. In fact, over 10,000 metric tons of finished plastic products are imported annually into Ghana. It is known that plastics do decompose, though not fully, over a very long period of time, on average about 100 to 500 years. That is to say plastics do not easily break down into simpler components or decompose. Commercially available plastics and polyolefins such as polyethylene, polypropylene, etc., are known to be resistant to decomposition by means of additional stabilizers such as antioxidants.

There is no denying the fact Ghanaians are not taking good care of the land. There has been serious erosion experienced in parts of the country partly caused by poor drainage. Stagnant water also comes with side effects like breeding moss, mildew, and insects that cause skin problems. Mining of the nation's mineral resources has been a public issue for a long time, not forgetting the environmental problem caused by galamsey. Sand winning is also a type of open-cast mining that provides material for the construction sector in Ghana, but the indiscrete winning of sand is causing more harm than good. The process of sand mining has accelerated environmental degradation to an alarming rate in many areas, and is accordingly putting pressure on vast farmlands and land for other communal purposes.

Ghana is also experiencing erratic rainfall and drier periods than before. The nation's originally 70 percent water cover in the 1980s has seen a massive

30 percent reduction over the last 30 years. Many residents cannot enjoy clean water for home use. All these are clear indications of climate change.

Theologians cannot sit by unconcerned when the lives of the people created in the image of God are being affected by environmental problems. It is believed that cleanliness is next to godliness. Theology must aim at promoting life in its fullness, hence one cannot underestimate the relevance of ecotheology. Let us now turn to a pragmatic interpretation of the creation account in Genesis 2:4b–17 from an Akan perspective to show how the Bible treats creation as essential to life and informs practical adherence to land care.

Creation Story as a Life Concern (Genesis 2:4b–17)

The Genesis creation stories highlight the beginnings of creation. Genesis 2:4b–17, which is the second creation story, focuses on the place and identity of humanity in creation as well as its mandate and role in creation. The setting of the identity is before the rain falling to water the land, the shrubs appearing, plants springing up. Human beings were created from the dust of the ground before the life-giving principle from God entered into the nostrils to give life (Gen 2:7). After that God planted a garden. Four rivers are mentioned to exist at that time flowing in the garden: Pishon, Gihon, Tigris, and Euphrates (Gen 2:10–14). These rivers watered the surface of the ground, presenting an idea that the presence of waterbodies is primarily essential to life in the garden. Life is made possible as long as rivers exist and flow to water the ground.

The command for the human being to keep the garden included the waterbodies which were flowing on the land. "Keeping" echoes the idea of guarding. Henry Morris avers that it is not about "protecting it from external enemies, of which there were none, but rather that of exercising a careful and loving stewardship over it, keeping it beautiful and orderly, with every component in place and in harmonious relationship with the whole."[24] The creation of humanity appears to precede the creation of the garden but not the land. Hence, the nature of the human person as being related to the land fundamentally defines identity. The Akan share a similar view when they say, *asaase a yete so yi, yedan no* (literally "the earth that we live on, we owe our existence to it"). The pun or word play by the writer by introducing *'et-hā'ādām* ("the man"; v. 7) with *'āpār min-hā'adāmāh* ("dust from the ground"; v. 7) draws home the interrelatedness between land and humanity. The pun can well be captured as the human being is "ground from the dust

---

24. Morris, *Genesis Record* 92–93.

of the ground" or the "earthling from the dust of the earth."²⁵ The well-being of humanity is closely linked to the well-being of the land.

The habitat of humanity is located in the *śedeh*, or "field" (v. 15). A serene field with trees and waterbodies is the ideal place for life. It foreshadows the writer's interest in humanity as an agent for managing the vast resources and products on the field. At this point in time, the field is yet to be productive. It needs rains from God to make vegetation grow. This means that God has a part to play in the development of the field. Thus, humanity's responsibility should be in tandem with God who brings rain, who is caring for the field in order to ensure proper development and growth. As long as the fields host waterbodies, animals, birds, and green vegetation, etc., human beings must collaborate with God in caring for what is there.

The task of humanity to *'abād*, or "till," the land denotes several ideas, critical among which is worship. The term *'abād* means "to serve, work, honor, worship." It connotes a responsibility and service to the object; that is to say, humanity has a responsibility to serve the ground. By serving the land, humanity serves God. Christian worship is service not only to God, but also to creation. According to Ralph Klein, "the Hebrew Bible actually has a verb that is translated literally as "serve it." In this understanding, humans are not so much over creation as under it. Ours is a servant role, making sure that the earth, its environment, its natural resources, its plants and animals survive, thrive, and grow."²⁶ Significantly, the notion of worship is embedded in servanthood. As humanity serves the land and tills it, worship is made complete. Disregarding and disrespecting the land will disrupt worship to God. The essence of keeping the land in its good shape should be a key factor in worship, and particularly realigned to a more environmentally-conscious vision of Christian responsibility.²⁷

It is God who "took" and "placed" humanity in the garden (Gen 2:15), echoing God's authority to set humanity to task where God desires. Humanity should not live in a vacuum, but in a context where attention on the land is concrete. It is in such a service that humanity can be accountable to God. Basically, humanity is seen as a steward in essence and also a servant of the ground. Walter Kaiser and D. G. Little's statement thus comes out forcefully: "for without a cosmos and an earth, there was not a story, or anything to talk

---

25. Kaiser and Little, *Biblical Portraits of Creation*, 29.

26. Klein, "Stewardship in the Old Testament," 334.

27. Bauckham, *Bible and Ecology*, 22; Osborn, *Guardians of Creation*, 86; Granberg-Michaelson, "Introduction," 1–5.

about, much less to live for."[28] Human identity, thus is closely linked to the whole of creation.

Implications for Theological Education

From the above discussions, a few inferences may be made. First, theological education in Africa must focus on promoting African identity, an identity that includes the relationship with the whole of creation. I stand for an inclusive African ecotheology, a topic that is included in all disciplines. This is because Western ecotheology has been skewed to alienate the individual from community and nature. Such an African ecotheology must focus on humanity's relationship with nature from an African perspective so that it can go a long way to define African identity. In fact, African identity is not in individual histories which have always projected supremacy, alienation, and preferability. It has little to do with race, color, or ethnic backgrounds. It is not simply about notions of slavery, liberation, or nationalism. It is not just in African art, traditions, or rites. It is not necessarily about notions of development, whereby one can claim to be developed and the other underdeveloped. It is not in plurality of ideologies and orientations. It is not essentially in psychodynamic traditions where a person's identity is an assimilation of external egos or in connection between community that develops a psychic structure as in Freud's or Erikson's theories. It is not principally in sociological elements of identity either. That is to say, a single discipline cannot exhaust African identity.

African identity should be established in what was in the beginning and certainly beyond constructs. It is not what one becomes or who one becomes, but what one is. It must be located in a relationship with ecology. As such, by being cognizant of the diffusion of histories, race, developments, ideologies, traditions, etc., theological education in African can equip students to have an identity located in a holistic view of life so as to maintain "new relationships with the entire creation in an attempt to avoid destruction and preserve life for all creatures."[29] African theology must be liberated from being overwhelmed with constructs that distort African identity to established paradigms and narratives of inclusiveness that affirm the relationship with nature as a gift and responsibility from God. It is in this light that African ecotheology will reflect on behavioral change in terms of how to care for the world. As Daniel Salas observes:

---

28. Kaiser and Little, *Biblical Portraits of Creation*, 15.
29. Daneel, "African Independent Churches," 250.

Approaching issues of environmental degradation and justice from a solely religious perspective may not solve the crisis on its own, but religion has an unparalleled influence in shaping morals. Faith can urge humans to examine our existence and to determine methods of change and growth as a means of protecting our common home.[30]

Second, African ecotheology must focus on the mandate of God for humanity not only to save souls, but to promote life in abundance. Issues about life cannot be devoid of ecotheological concerns for if the last tree dies, the last man dies. Beatus Kitururu says, "Human growth is always interactive and dialogical. Human beings are interdependent and interconnected. It is with others that one can appreciate that one's potential life is relational."[31]

Third, a study of African ecotheology has everything to do with stewardship and worship. Stewardship goes beyond individual gifts to encompass the whole of creation. It seems the NT emphasizes more on the personal stewardship of the gifts God has bestowed on an individual (1 Pet 4:10). The parable of the talents espoused by Jesus reinforces such a view and makes it sound like caring for talents entrusted to an individual is what is required (Matt 25:14–30; Luke 12:42–43). As stewards, humanity must learn to do things the way God wants, and as Celia Deane-Drummond puts it, a "violation of the earth amounts to an affront to God's holiness."[32] In his *Wisdom in Israel*, Gerhard von Rad pointed out that Israel had to pay heed to the voice of nature that calls them to treat it with respect. For the sages of Israel, the natural order was not an object to be manipulated: "The man who has brought his relationship with God into order is himself in league with the stones of the field and is befriended by the beast of the field (Job 5.20ff.). The environment is not only the object of man's search for knowledge. In so far as it is turned towards him, man is the object of its advance on him."[33] Care for all creation should be seen as an act of worship that attracts a blessing. It is also in this context that the worshipper can see the power and goodness of God through nature (Rom 1:20). Focusing on the mandate God has given will help create a cosmic balance in knowing that God wants humanity's response to be including actions that brings hope to one's environment.

---

30. Salas, "Consumption and Crisis," 80.
31. Kitururu, *Spirituality of Hospitality*, 4.
32. Deane-Drummond, *Eco Theology*, 89.
33. von Rad, *Wisdom in Israel*, 302.

Fourth, promoting African ecotheology will enhance spiritual growth. African spirituality is not only a vertical adjustment to God but also a horizontal experience with other persons and nature. African spirituality finds expression in how people live and calls for an actively interactive relationship. It is about how human beings relate to the complex world where the spiritual and physical merge, the life-bearing and lifeless aesthetics interact, and the animate and inanimate powers become responsible and participative to the vital life of nature. According to McCarthy, authentic spirituality can never be individualized or exclusive; it participates in the joys and sorrows of the real world and embraces actions that aim at repairing the world and its brokenness.[34]

Spirituality that focuses on one's personal growth is alien to African life and thought. Peter Paris likens African spirituality to the variations of tones that make music, and says, "Unity in diversity is another metaphor for African Spirituality."[35] Furthermore M. A. Umeagudosa writes: "Spirituality is an embodied, lived experience, influenced by historical, cultural and social circumstances. It includes cognitive orientation, it transcends the intellect and is formed by and gives form to our relations, experience of our bodies, our communities, our ecstacies and our aesthetics."[36] Tendencies that restrict spirituality to oneself are selfish, and as Kitururu observes, "It is not surprising, therefore, to realize that indigenous spirituality holds that evil is inherent in human selfishness."[37] The call is on humanity to recognize and affirm authentic values of life present in African cultures of communion, philosophies of togetherness, and spirituality of connectedness that thrives on reconciliation between nature and humanity. In fact, Leonardo Boff asserts that the cry of the earth and the cry of the poor are inseparable.[38] No wonder during festivals in some parts of Africa, there are days set aside to pacify and reconcile with nature. African Christian spirituality therefore must find expression in ways in which Africans relate with God and nature by affirming the authentic values of life while exploring the creative tension between the word of God and what gives meaning to life in its fullness.

Finally, theological education should equip Africans to stand out in African government's efforts to regulate environmental risks associated with land care, including pollution of the land, oceans, waterbodies, and wildlife. African theologians must know what to say and contribute to the

---

34. McCarthy, "Spirituality in a Postmodern Era," 200.

35. Paris, *Spirituality of African Peoples*, 22.

36. Umeagudosa, "Primacy of Prayer," 23.

37. Kitururu, *Spirituality of Hospitality*, 33.

38. Boff, *Cry of the Earth*, 114.

agenda of land care in the public sphere and must carve an authentic identity, holding on to the mandate God has given, so that it can affect lives and society positively. It will then be that society will know the impact of theological education in real life.

Conclusion

The object of this paper has been to advocate for theological study in Africa that reflects an integrated ecotheological agenda, and which equips theological students to be partners in caring for land, water bodies, and the environment. It has proposed matters of African identity, the biblical mandate of nature's care especially in Genesis 2:4b–17, and an authentic African spirituality to feature in every theological and religious discipline and curricula as part of African ecotheology. It has held that when attention is placed on ecotheological issues, students can develop a voice to participate in social and public discourse on nature care and can direct theological reflections on human vocation and care for the land. Access to education has tremendous impact on people, helping them to be aware of their rights and of how to make rational decisions. It empowers people to protect themselves from abuse and oppression. It prepares people for the job market and helps to alleviate poverty. Theological institutions need to equip themselves to stand out in the African government's efforts to regulate environmental risks associated with land care, including pollution of our oceans, water, and wildlife. Such an agenda can influence the church in its ecclesial mission and responsibility to the social and public life of earth's care. Theological education in Africa must focus on how one's personal affective development will influence others and transform society and nature.

Bibliography

Bauckham, Richard. *The Bible and Ecology: Rediscovering the Community of Creation*. London: Baylor University Press, 2010.
Boff, Leonardo. *Cry of the Earth, Cry of the Poor*. Translated by Philip Berryman. Maryknoll, NY: Orbis, 1997.
Brueggemann, Walter. *The Land: Place as Gift, Promise, and Challenge in Biblical Faith*, Overtures to Biblical Theology 1. Philadelphia: Fortress, 1977.
Conradie, Ernst M. "Contemporary Challenges to Christian Ecotheology: Some Reflections on the State of the Debate after Five Years." *Journal of Theology for Southern Africa* 147 (2013) 108–16.
Crane, Michael "Equipping the Transient for Ministry in a Global City." *The New Urban World Journal* 3.1 (2014) 2–18.

Daneel, M. L. "African Independent Churches Face the Challenge of Environmental Ethics." In *Ecotheology: Voices from South and North*, edited by David G. Hallman, 248–63. Maryknoll: Orbis, 1994.

Deane-Drummond, Celia. *Eco Theology*. London: DLT, 2008.

Fretheim, Terence E. "Preaching Creation: Genesis 1–2." *Word and World* 29.1 (2009) 76–79.

Granberg-Michaelson, Wesley. "Introduction: Identification or Mastery?" In *Tending the Garden*, edited by Wesley Granberg-Michaelson, 1–7. Grand Rapids: Eerdmans, 1987.

Kitururu, Beatus B. A. *The Spirituality of Hospitality: African and New Testament Perspectives* Nairobi: Catholic University of Eastern Africa Press, 2009.

Klein, Ralph W. "Stewardship in the Old Testament." *Currents in Theology and Mission* 36.5 (October 2009) 330–34.

Krathwohl, David, et al. *Taxonomy of Education Objectives Handbook II: The Affective Domain*. New York: David McKay, 1964.

Light, Andrew. "Urban Ecological Citizenship." *Journal of Social Philosophy* 34.1 (2003) 44–63.

Martin, Barbara L., and Charles M. Reigeluth. "Affective Education and the Affective Domain: Implications for Instructional-Design Theories and Models." In *Instructional-Design Theories and Models: A New Paradigm of Instructional Theory*, edited by Charles M. Reigeluth, 485–509. Mahwah, NJ: Lawrence Erlbaum, 2001.

Marx, Karl. *The German Ideology*. Abridged ed. Edited by C. J. Arthur. New York: International, 1970.

McCarthy, Marie. "Spirituality in a Postmodern Era." In *The Blackwell Reader in Pastoral and Practical Theology*, edited by James Woodward and Stephen Pattison, 192–204. Oxford: Blackwell, 2000.

Moltmann, Jurgen. *Theology of Hope*. Minneapolis: Fortress, 1993.

Morris, Henry M. *The Genesis Record: A Scientific and Devotional Commentary on the Book of Beginnings*. Philadelphia: Baker, 1976.

Nelson, Michael Paul, and Leslie A. Ryan. "Environmental Ethics." DOI: 10.1093/OBO/9780199363445-0025

Oduro, Thomas A. "Theological Education and Training: Challenges." *Journal of African Instituted Church Theology* 2.1 (2006) 4–16.

Okure, Teresa. "Feminist Interpretations in Africa." In *Searching Scripture: A Feminist Introduction*, Volume 1, edited by Elisabeth Schüssler Fiorenza, 76–81. London: SCM, 1993.

Osborn, Lawrence. *Guardians of Creation: Nature in Theology and the Christian Life*. Leicester: Apollos, 1993.

Paris, Peter J. *The Spirituality of African Peoples: The Search for a Common Moral Discourse*. Minneapolis: Fortress, 1995.

Pearson, Clive. "Electing to do Ecotheology." *Ecotheology* 9.1 (2004) 7–28.

Rad, Gerhard von. *Wisdom in Israel*. Translated by J. D. Martin. London: SCM, 1972.

Santmire, Paul. *The Travail of Nature: The Ambiguous Ecological Promise of the Christian Faith*. Minneapolis: Fortress, 1985.

Sills, David. *Hearts, Heads, and Hands: A Manual for Teaching Others to Teach Others*. Nashville: Broadman and Holman, 2016.

Smit, Dirk J. "Modernity and Theological Education—Crises at 'Western Cape' and 'Stellenbosch?'" *Journal of African Christian Thought* 2.1 (June 1999) 33–44.

Umeagudosa, Margaret Azuka. "The Primacy of Prayer in African Spirituality: Tradition and Christian." In *African Dilemma: A Cry for Life*, edited by Kofi Appiah-Kubi, 21–35. Nairobi: EATWOT, 1991.

Werner, Dietrich. "Introduction." In *Eco-Theology, Climate Justice and Food Security: Theological Education and Christian Leadership Development*, edited by Dietrich Werner and Elizabeth Jeglitzka, 13–18 (Geneva: Globethics.net Global, 2016).

The Bible and Caring for the Land: A Theological Approach

Thomas A. Oduro
Good News Theological Seminary, Oyibi, Accra

Abstract

The destruction of river bodies, forests, mountains, and other environmental facets in Ghana, a country where over 70 percent of the population is Christian, is quite paradoxical. The sad state of the environment and its potential consequences on posterity, if not checked, will lead to a total destruction of the ecosystem. This article throws a searchlight on the importance of keeping a sacred environment, vis à vis the apathetic attitude of Christians and Christian institutions toward environmental degradation in Ghana. Lastly, a biblical paradigmatic theological approach meant to protect the environment has been proposed.

Introduction

THE FIRST TWO CHAPTERS and other verses in the Bible unequivocally state that everything, seen and unseen, was created by God. Thus, God is the owner of whatever is in the world, seen and unseen. God perceived whatever he created as good (Gen 1:1—2:15; Pss 24:1, 2; 89:11). God appointed Adam, the one from whom all human beings descend, as the steward of the created order (Gen 1:26–31). He was commanded to "subdue" and "rule" the earth. Adam was not only appointed a steward, he was made dependent on God's created order. Human beings, as a result, have become dependent on nature since creation. Christians, without any ambiguity, believe that "The earth is the Lord's and the fullness thereof, the world and those who dwell in it," (Ps 24:1) yet how to balance contingency and the ruling of the environment has become paradoxical.

Traditional Religious Approach

Unlike Christians, adherents of African traditional religion neither have a sacred book nor a clear common belief about the genesis and ownership of the environment and the role human beings are expected to play to keep the environment thriving. Nevertheless, the ancestors of the adherents of traditional religions had a knowledge of human beings' stewardship of the contingency on the land.[1] The Akan, for instance, regard the earth as a spirit of God; so, they have personified the earth as a female deity with a name, *Asase Yaa*.[2] Therefore, Thursdays are set aside for the land to rest, so no work is done on that day.[3] The adherents of traditional religions unequivocally hold on to the total dependency of humanity on the female deity—earth:

> 'Earth, whether I am alive or dead, I depend on you.' By this you [the ancestors] did not only mean that we walk on the earth and when we die we shall be buried in the earth. You meant something more than that. You were referring to the fact that whatever keeps us in existence is derived from the earth. It is the earth that gives us the water that we drink to live; the earth produces the food without which [we] would be dead; the earth furnishes us with the trees with which we build our houses. We depend upon the earth to play, to travel, to hunt, to perform our rituals and ceremonies. The earth, indeed, is a mother to us. Its fertility is our live-wire. It is for this reason that you [the ancestors] did everything to keep the sacredness of the earth intact.[4]

Adherents of traditional religions did not only personify the earth, but every component of it, as stated above: water bodies, fish in the water bodies, food, trees, pathways, animals, birds, and insects. "Almost all West African communities believe that there are deities inhabiting the waters,

---

1. "Land," "earth," and "environment" are used interchangeably in this article.

2. *Yaa* is a name for females born on Thursday.

3. Opoku, *West African Traditional Religion*, 56. The Earth is known among the Fante as *Asase Efua*. No work is done on it on Friday, a day believed to be her soul day.

4. Sarpong, *Dear Nana*, 29.

great and small. They are looked upon as beneficent deities who preside over the sea, rivers, lakes and lagoons."[5]

Deep inside the bowels of the earth in Ghana are precious minerals such as gold, bauxite, diamond, manganese, iron ore, aluminum, cement, lithium, limestones, etc. The ancestors, who did not have the technical expertise to mine the minerals, left the mining of the riches of the land to posterity. They, however, kept a healthy balance of stewardship and contingency on the land by declaring some rivers, trees, fish, and animals as sacred beings, which are believed to be either inhabited by the gods or are objects of the gods. Some forests and groves were declared sacred; holidays were assigned to them. Rituals were performed at such sacred places; libations were poured to appeal to the gods for sustained provision, protection against enemies, and the aversion of calamities. Myths and taboos about the sacred places kept people from destroying the sanctity of the sacred objects:

> In most Akan and Ga communities there were days which were considered bad days for farming, fishing, and hunting. It was believed the gods will strike one dead if he/she did. One will not return if one ventured into the forest, farm or sea. Our elders constructed this taboo to preserve the environment and its inhabitants. If one hunts or fishes each day, then one day the animals in the forest will be extinct and all the fishes in the sea will be gone. If one day was preserved then at least they can be protected. Sometimes even a whole season was reserved so the fingerlings and infants in the forest could grow and also multiply. During this period no one dared to venture into the forest, farm or sea. This served as a fallow period to preserve the animals and environment.[6]

Those who flouted the rules regarding the sanctity of the environmental objects were perceived as enemies of the community; they were, therefore, severely punished. Adherents of traditional religions were particularly concerned about God's presence in the hills, mountains, rivers, lakes, groves, etc. They are amazed by God's mystery in such creation, so they worship them. The attitude of the adherents of traditional religions to the created order, though pantheistic, dignifies God's creation by keeping the environment neat, sacred, sustainable, and habitable for human beings, animals, birds, and insects. Rivers and other water bodies were drinkable, thus, serving as ecosystems that made life worth living.

---

5. Opoku, *West African Traditional Religion*, 60.
6. Nyarko, "Spirit of Taboos."

The Christian Factor

The advent of Christianity in Africa, south of the Sahara, by Western missionaries from the late fifteenth century onward, caused devastation in the subduing and ruling of the earth by Africans. African Christians who were appalled by the rituals that went with the preservation of the land perceived the concept of keeping the land holy and productive by the adherents of traditional religions as an act of worshipping the creation instead of the Creator. As a result, the regulations that kept the land sacred were recklessly broken by some Christians. The sad state of the environment in Ghana is a testimony of Christian complicity in the degradation.[7]

Theology has been defined in many ways by many scholars. Etymologically, the two Greek words *theos,* meaning "God," and *logos,* meaning "word" or "reasoned discourse," has put the definition of theology ajar.[8] Charles Hodge defines it as "the science of the facts of divine revelation so far as those facts concern the nature of God and our relation to him, as his creatures, as sinners, and as the subjects of redemption."[9] To John Parratt, theology is "our apprehension of God based on His self-revelation."[10] Clemens Sedmak defines theology simply as "talking to people about God, or talking about God in the light of God's presence in the world."[11] The definitions influence many Christians to perceive theology as only a discourse about God in the lives of humanity, nations, and what happens in the world based on God's self-revelations. Consequently, many Christians do not consider the earth and the other created order as part of God's self-revelation. Hence, in most cases, theological discourse excludes the earth and all that there is. John Parratt, however, corrects that notion by stating that "All our [i.e., Christians] thinking about God has to take into account the assumption that the world is God's creation. Our theology will thus have to be large and comprehensive enough to embrace and relate to all aspects of the natural and human sciences—but at the same time without losing its particular convictions and insights."[12]

The Bible does not only reveal the created order as mysterious, it reveals God's presence to human beings: "The heavens declare the glory of God, and the sky above proclaims his handiwork" (Ps 19:1). Christian

---

7. Korasare "An Open Letter."
8. Parratt, *Guide to Doing Theology,* 2.
9. Hodge, *Systematic Theology,* 21.
10. Parratt, *Guide to Doing Theology,* 3.
11. Sedmak, *Doing Local Theology,* 6.
12. Parratt, *Guide to Doing Theology,* 2.

apologetics use the created order to prove the existence of God.[13] Seminary students are taught all kinds of theologies—biblical theology, historical theology, sacramental theology, systematic theology, contextual theology, feminist theology, moral theology, confessional theology, etc. Hardly ever, however, are they taught a theology of the earth,[14] mostly because many Christians are amazed at who God is more than at his creation of the earth and all there is in it. Many Christians undoubtedly and intellectually know that "The earth is the Lord's and the fullness thereof, the world and those who dwell therein" (Ps 24:1; ESV), but they fail, pragmatically, to subdue the earth as God's property.

Theology of Mountains

If theology is having a discourse of God based on his self-revelation to humanity, then the discourse must encapsulate God's revelation to humanity in not only creating the earth and all there is in it, but also how God identifies himself with the created order in the Bible. An example of using mountains as a theological approach to restore the dignity of the ecosystem is postulated. Mountains, in biblical times, were prominently linked with God's presence. God specifically commanded Abraham to sacrifice Isaac on one of the mountains in the land of Moriah (Gen 22:1, 2) so Abraham called that mountain "the mount of the Lord" (Gen 22:14). Each of the mountains—Horeb and Sinai—was designated as the "mountain of God" (Exod 3:1; 4:27; 19:11; 24:13). In fact, the Israelites perceived Mount Sinai as literally being the dwelling place of God on earth because of the numerous times Moses communed with God on it. One wonders why God chose to reveal himself on a mountain. Again, one also wonders what the attitude of the Israelites toward Mount Sinai was even when nothing peculiar was happening. No wonder mountains formed part of the theology of Israel. To the psalmist, those who trust in the Lord are, metaphorically, as immovable as Mount Zion, while the protective power of God toward his people is likened to the impregnability of the mountains that surround Jerusalem (Ps 125:1, 2). With this theology, the Israelites no longer perceived mountains as "a landmass that projects conspicuously above its surroundings and is higher

---

13. For more knowledge on the cosmological argument to prove the existence of God, read Oden, *Classic Christianity*, 80–94.

14. The writer looked at the curriculum of Trinity Theological Seminary, Accra, The Good News Theological Seminary, in Accra, and Tyndale Theological Seminary, the Netherlands.

than a hill."[15] Many mountains were perceived as sacred objects, a perception that determined society's approach to them.

The theology of mountains continues in the New Testament. Jesus Christ used some elements of the earth to teach many truths. He used a mustard seed and a mountain as figures of speech to illustrate the immensity of divine authority that is embedded in his believers to achieve what is imaginably impossible. "And he said to them, 'Because of the littleness of your faith; for truly I say to you, if you have faith as a mustard seed, you shall say to the mountain, "Move from here to there," and it shall move, and nothing shall be impossible to you'" (Matt 7:20; 21:21). The reason why the transfiguration of Jesus—an event that was reminiscent of the divine presence on Mount Sinai—took place at no other place but on a high mountain (Matt 17:1–8) reinforces the significance of the theology of mountains.

Despite the unambiguity of the theology of mountains in the Bible and the fact that there are more than twenty-one mountains[16] and mountain ranges in Ghana, one rarely hears any sermon preached or gospel song sang about Ghanaian mountains. Some Christian congregations, however, pride themselves by using the names of some biblical mountains to prefix their names. As such, there are names like Mount Zion Methodist Church, Mount Olivet Methodist Church, Mount Zion Assembly of God, Mount Moriah Presbyterian Church, and Mount Zion Presbyterian Church. The sad part of it is that most of the members have not seen the mountains that define the identity of their congregations. The nonformulation of a theology of mountains in Ghana has made Ghanaian Christians disregard the essence of the land and its other elements.

Theology of Water Bodies

The River Euphrates is mentioned from the beginning to the end of the Bible (Gen 2:14; Rev 9:14). It is so prominent that it is simply called "the river" in many parts of the Bible. Rivers in the Bible were held in high esteem due to their usefulness to humanity. Many of the rivers of the Bible are still in existence, such as the Euphrates, Jordan, and Nile. As a result, a theology was developed and articulated about rivers and streams. To this end, Isaiah prophesied that God will extend peace to Israel like a river, and the glory of the nations like an overflowing stream (Isa 66:2). Amos the

---

15. Meriam-Webster, *Meriam-Webster's Collegiate Dictionary*, Tenth Edition.

16. See https://peakery.com/region/ghana-mountains. Some of the tallest mountains are Afadjato, Edouka, Atiwiredu, Kwamisa, Pakesie Bepo, Atwea, and Aburi. See Gracia, https://yen.com.gh/109470-names-mountains-ghana-locations.htm#109470.

prophet admonished leaders of Israel to "let justice roll down like a river and righteousness like an ever-flowing stream" (Amos 5:24). The use of the simile *like* indicates the people's knowledge and awareness about the flow of rivers and streams. It also indicates the unstoppable trend of God's peace. Similarly, leaders of Israel were not left in doubt about the divine expectation of justice in society. The psalmist illustrates the immense authority of God by stating that God was going to change rivers into a wilderness and springs of water into a thirsty ground (Ps 107:33). Isaiah prophesied that God will open rivers on the bare heights, and springs in the valleys: "I will make the wilderness a pool of water and the dry land fountains of water" (Isa 41:18). The omnipotence of God was clearly illustrated with the theology of rivers, streams, and the wilderness. The audience of these utterances, therefore, did not have any difficulty in understanding the implications of theology since the water elements were known to them.

The Present Reality in Ghana

The degradation of the environment in Ghana is so obvious that many people, including church leaders, have expressed grave concern. During the official opening of the 38th Annual Synod of the Sunyani Diocese of the Methodist Church, Ghana, at the Emmanuel Methodist Church, Nkwabeng, a suburb of Sunyani in the Brong Ahafo Region, the Right Reverend Kofi Asare Bediako, the Bishop of the Sunyani Diocese of the Methodist Church, Ghana, expressed worry at the fast rate at which the Ghanaian environment is degrading. The Right Reverend Asare Bediako cited illegal mining, contamination, and destruction of water bodies, the menace of plastic waste and littering, open defecation, indiscriminate burning of vegetation, illegal lumbering, and lack of reforestation as some of the human activities that had affected the environment and the quality of lives of the people.[17]

Nana Addo Dankwa Akufo-Addo, the President of the Republic of Ghana, speaking at the opening session of the 2018 West African Mining and Power Conference and Exhibition (WAMPOC/WAMPEX) on Wednesday, 30 May, in Accra, lamented how mining has degraded the living conditions of the inhabitants of four towns in Ghana where gold and diamonds are mined:

> Why is Obuasi not the most beautiful city in Ghana or the world if it hosts the richest gold mine? Why do Tarkwa and Prestea not look like the golden towns they are? And why does Akwatia's

17. "Bishop Expresses Worry."

appearance not reflect anything about the diamonds that have been taken from its soils all these years? . . . The distressed state of communities in which mining companies operate is nothing short of a disgrace and we must work to change the situation.[18]

With the destruction of the land, rivers, streams, and other water bodies in Ghana, it is difficult to formulate a meaningful environmental theology for Christians to comprehend because many Christians are at the forefront of destroying the environment, God's gifts to Ghanaians.[19] Paradoxically, the River Jordan has been so preserved that water from it is sold to pilgrims to Israel from many countries.[20] Had such rivers and streams been destroyed in Israel, pilgrims would not have had the opportunity to see them, let alone take some to Ghana.

Glimpses of Environmental Theology in Ghana

Despite the apparent apathy of some Christians in Ghana to develop theologies that would enable Ghanaians to appreciate the earth and everything in it, only a few Christians are doing their best to formulate environmental theologies. Afua Kuma, an illiterate farmer in the forests of Kwahu, uses the forest, among other things, to expatiate Akan Christological theology. Jesus is:

> The great rock we hide behind:
> the great forest canopy that gives cool shade:
> the Big Tree which lifts its vines
> to peep at the heavens,
> the magnificent Tree whose dripping leaves
> encourage the luxuriant growth below.[21]

Afia Kuma uses land and sea to illustrate the authority of Jesus:

> You weave the streams like a plaited hair;
>> with fountains you tie a knot.
>> Magician who walks on the sea:
>> he arrives at the middle,
>> plunges his hand into the deep and takes out a whale.[22]

18. "Mining Companies Not to Blame," paras. 4–5, 8.
19. See "Christians are Both."
20. "Holy Land Mall," para. 1.
21. Kuma, *Jesus of the Deep Forest*, 5.
22. Kuma, *Jesus of the Deep Forest*, 6.

To those who search for someone to depend on to fulfill a dream or achieve a goal, Afua Kuma directs that "Jesus is the grinding stone on which we sharpen our cutlasses."[23] Afua Kuma's Christology, which is backed by elements of the earth and sky, is one of the finest ways of developing a local theology that is easily understandable to people who are in a particular context. It is a clear paradigm of an environmental theology in Ghana.

Christian prayer camps, forests, gardens, and grottos have become common phenomena in Ghana for the past three decades. They are designated areas where people, who perceive themselves as needing something special and urgent from God, rush to pray and fast, often assisted by the leader and assistants of the camps. Meditation is usually the main activity of those who go to the grottos. The seven most patronized prayer camps are: Moments of Glory Prayer Army (MOGPA) in Kumasi, Abasua Prayer Camp located on Atwea Mountain in Ashanti Region, Bethel Prayer camp in Kumasi, Achimota Forest in Accra, Mount Horeb Prayer camp at Mamfe, Hebron Prayer camp at Nsawam, and Resurrection Power New Generation Church Prayer camp in Accra.[24] Although the prayer camps, gardens, grottos, and forests are serene by nature, one wonders what actually draws the people to patronize the areas. Is it due to the serenity, divine presence, or the spiritual prowess of the owners of the camps and the grottos? No matter what prompts people to patronize the camps, they are places that could be used to articulate an environmental theology befitting to God.

A Call to Doing an Environmental Theology

Theologians are generally perceived to be those who have studied and specialized in detail aspects of Christian doctrine or practice. They are perceived as those who hold degrees in specialized area of study. Stone and Duke, however, argue that "All Christians are theologians. It's not that they were born that way or decided one day to go into theology. It's a simple fact of Christian life: their faith makes them theologians, whether they know it or not, and it calls them to become the best theologians they can be."[25] This concept does not exclude anyone from doing theology. Every Christian is

---

23. Kuma, *Jesus of the Deep Forest*, 17.
24. https://buzzghana.com/largest-prayer-camps-ghana-now/. MOGPA is headed by Rev. Osei Bonsu. Abasua was established by Rev. Abraham Osei-Asibey, a Methodist pastor. Hebron Prayer camp is headed by Elder Donkor of the Church of Pentecost. For more information on Prayer Camps, see https://yen.com.gh/57819-here-top-5-prayer-camps-ghana-jesus-christ-stays.html#57819.
25. Stone and Duke, *How to Think Theologically*, 1.

called to do theology. As such, the scope of theology is expansive and limitless; far beyond the imagination of most Christians:

> Every aspect of the life of the church and its members is a theological testimony. So too are the particular ways Christians have of relating to what is around them, their styles of interacting with others and the world. Even if they were to run off and live in some remote . . . forest . . . that decision would itself be a theological one. Likewise, choices they make about their wilderness lifestyle—how to live among the trees, the lakes, and the animals—would signal something of their understanding of God's creation. *To be Christian at all is to be a theologian. There are no exceptions.*[26]

Chip Ingram adds his voice to the theological roles of Christians when he says:

> Someone taught me early in my journey, "A disciple always leaves things better than he found them." We need to recognize how our heavenly Father feels and thinks about this planet He placed under our dominion. We have power to subdue but not abuse God's possession. We are to cultivate and care for it because we are vice-regents over the earth.[27]

Conclusion

Revitalizing the earth and its elements that have been destroyed in Ghana is not the singular duty of the mining companies, government agencies, security services, task forces, NonGovernmental Agencies, or volunteer groups. Every Christian is called to play a theological role in the forms of songs (both composed and sung), movies, paintings, novels, poetry, drama, drum language, sculpting, preaching, teaching, and engaging in other pragmatic roles such as debates, public lectures, street marches, inscriptions on buses/motor cars, and the writing of books that advocate the preservation of the earth whilst depending on it for sustenance. Students studying theology in seminaries must, as a matter of policy, be mandated to develop a theology on the need to subdue and rule the earth responsibly as a pre-graduation requirement. When Christians in Ghana, who constitute over 70 percent of the population, exemplify what God expects from them, theologically,

---

26. Stone and Duke, *How to Think Theologically*, 2. Emphasis not mine.
27. Ingram, *Culture Shock*, 173.

about responsible stewardship of the earth and all there is in it and about it, the environment in Ghana will reflect the expectations of God.

Bibliography

"Bishop expresses worry over environmental degradation." https://www.ghanabusiness news.com/2017/04/18/bishop-expresses-worry-over-environmental-degradation/

"Christians Are Both the Cause of and Solution to the Galamsey Menace in Ghana." http://m.peacefmonline.com/pages/comment/features/201704/310794.php.

Gracia, Zindzy. https://yen.com.gh/109470-names-mountains-ghana-locations.htm#109470.

Hodge, Charles. *Systematic Theology*, Vol. 1. Grand Rapids: Eerdmans, 1993.

"Holy Land Mall." http://www.holylandmall.net/holwatfromjo.html.

Ingram, Chip. *Culture Shock: A Biblical Response to Today's Most Divisive Issues*. Grand Rapids: Baker, 2014.

Korasare, Elizabeth. "An Open Letter to the Christian Community of Ghana." https://www.graphic.com.gh/features/opinion/an-open-letter-to-the-christian-community-of-ghana.

Kuma, Afua. *Jesus of the Deep Forest: Prayers and Praises of Afua Kuma*. Translated by Jon Kirby. Accra, Ghana: Asempa, 1981.

Merriam-Webster Incorporated. *Merriam-Webster's Collegiate Dictionary*, Tenth Ed. Springfield, MA: Merriam-Webster, 1993.

"Mining Companies Not to Blame for Impoverished Communities – Mutawakilu." https://www.google.com/search?client=firefox-bd&q=%E2%80%9CMining+Companies+Not+to+Blame+for+Impoverished+Communities+%E2%80%93+Mutawakilu.%E2%80%9D

Nyarko, Daniel. "The Spirit of Taboos." http://www.ghanaculture.gov.gh/mod_print.php?archiveid=2297.

Oden, Thomas C. *Classic Christianity*. New York: HarperOne, 1992.

Opoku, Kofi Asare. *West African Traditional Religion*. Accra, Ghana: FEP International, 1978.

Parratt, John. *A Guide to Doing Theology*. London: SPCK, 1996.

Sarpong, Peter K. *Dear Nana: Letters to My Ancestor*. Takoradi, Ghana: Franciscan, 1998.

Sedmak, Clemens. *Doing Local Theology: A Guide for Artisans of a New Humanity*. Maryknoll, NY: Orbis, 2002

Stone, Harold, and James O. Duke. *How to Think Theologically*. Minneapolis: Fortress, 1996.

The Bible and Caring for the Land: African Theocology as Christian Impulsion for Creation Care

Ebenezer Yaw Blasu

Akrofi Christaller Institute of Theology,
Mission & Culture, Akropong

Abstract

The paper argues that an African Christian theology that can motivate moral responsibility for creation care needs to be an engagement of the Bible with African ecological thought and praxis. The result will be African Theocology—a way of studying ecological science from the perspective of God and his relations with his creation in the African context. Drawing instances from a study among the Sokpoe-Eve of Ghana, the paper deduces three elements in African ecological praxis which can be engaged with the Bible in reconstructing African Christian theocology. Furthermore, the axiological implications of both African anthropocentric and biblical theomorphic valuing of humanity can provide impulsion for us to care for creation as ecological vicegerents. African theocology then hopes to provide a theocentric Christian impulsion to care for creation in either religious or scientific ecology.

Introduction

IN MY VIEW, DISCUSSIONS about the Bible and caring for the land suggests the need for a Christian or Bible-based teaching or ideology that can propel and shape the political campaign against the destruction of land and the environment (the earth). It reminds me of Kwame Bediako's suggestion from which I infer that the Christian Scriptures need to be a worthy source and, hence, central in engaging the gospel with ecoculture in thinking of and reconstructing a theology for creation care.[1] For Bediako, Scripture (Bible) is not only the yardstick, the model and measure for testing, but also points to and urges or impels us for all our engagements in the continuing divine-human-nature encounter that characterizes Christian faith.[2] This paper shall propose and argue that African theocology is a synthesis of Christian religious and scientific ecology that promises to pragmatically provide impulsion for creation care. First, it shall argue for the role of religion in motivating for ecological actions; and then suggest three concepts in the worldviews and ecoculture of the Sokpoe-Eʋe of Ghana for engagement with Scripture in developing African theocology curricula.[3] The paper draws comparative relations between theocology and ecotheology, and proposes in the end that African theocology can be studied in the fields of both ecological science and religious ecology.

The Role of Religious Ecology in Caring for the Earth

John A. Grim and Evelyn Mary Tucker rightly observe that as the understanding of nature deepens and broadens, especially regarding the global ecocrisis, the legitimate question about our role in nature is being recognized and asked: "How do humans fit in ecosystems and what modes should guide our moral responsibility to restore ecosystems and why?"[4] Secular ethicists disenchant and reject any spiritual defense of nature because "it would seem more appropriate nowadays to find a secular 'and scientifically informed public' justification for government action to protect the

---

1. Bediako, "Scripture as the Hermeneutics," 2.
2. Bediako, "Scripture as the Hermeneutics," 3.
3. Blasu, "Christian Higher Education as Holistic Mission." Some students and lecturers of Environmental Science (GNSP 101) at Presbyterian University College, Ghana who participated in this study had problems integrating biblical faith in teaching and learning a subject designed purely as a science and yet expected to induce environmental moral responsibility.
4. Grim and Tucker, *Ecology and Religion*, 62.

environment."[5] Yet scientific discourse and education have not been and are not a panacea to resolving ecological crises.[6] Science is a contributor to the dominant anti-ecological culture,[7] as it does not promote responsible earthiness or rooting for creation care in us.[8] Dianne Bergant also implies in her *The Earth is the Lord's* that scientific and technological facts lead us to believe that we can step outside of our environment to examine it and control it.[9] From her perspective, science and technology tend to give a false sense of being apart from and not part of the environment, as if "we merely live within the environment as we live within a building."[10] The result is the lack of ethical impetus in scientific worldview and accounts for its inability to replenish the depleting material resources.[11] Ghillean Prance, former Director of the Royal Botanic Gardens at Kew, reasons out that the ecocrisis is a moral, spiritual, and ethical one requiring major changes in our behavior to complement science in creation care.[12] For their part, Grim and Tucker have shown that neither scientific facts nor civil regulations, but a "synthesis of religious ecology and ecological knowledge [science]," is what is required[13] for effective Christian environmentalism. Christians need to pursue ecological science and scholarship from a biblical perspective as an integral part of believers' total commitment to God[14] with a hope of moral transformation for creation care.

The inadequate scientific approaches may explain why religious ecology and the need for reconstructing theologies for creation care enjoy significant attention now. Religious ecology helps humans to understand and reenvision their roles as morally responsible participants in the dynamic processes of life in the ecosystem.[15] It implies discovering the meaning of creation and humankind's place within it from the perspectives of religious traditions.[16] John Grim and Mary Evelyn Tucker argue that because religious ecology "in the past sustained individuals and cultures in the face

---

5. Curry, *Ecological Ethics*, 137. See also Eckersley, "Beyond Racism," 178.
6. Bookless, *Planetwise*, 43.
7. Curry, *Ecological Ethics*, 137.
8. Bookless, *Planetwise*, 50.
9. Bergant, *Earth is the Lord's*, 10.
10. Bergant, *Earth is the Lord's*, 10.
11. Bergant, *Earth is the Lord's*, 10.
12. Ghillean Prance, cited in Bookless, *Planetwise*, 43.
13. Grim and Tucker, *Ecology and Religion*, 157.
14. Son, "Relevance of a Christian Approach," 2.
15. Grim and Tucker, *Ecology and Religion*, 63.
16. Özdemir, "Toward an Understanding," 5.

of internal and external ecological threats," there is "a growing consensus that religions may play a significant role now also."[17] In Ghana, for instance, Allison Howell called on both government and religious bodies, especially Christians, to return to religious engagement with the environment. This call was a response to her observation that "the spiritual engagement with land and water, once part of the fabric of African spirituality, seems to have become unravelled in our time."[18]

Arun Agrawal and C. C. Gibson, however, observe that there might be significant challenges with employing primal religious ecological ethics, good as they are in themselves, in contemporary times.[19] Ironically, like scientific ecology, religious ecothinking and practices are also not necessarily exonerated from contributing to the problem of global ecocrisis.[20] Nevertheless, they also are more essentially part of the potential way forward for creation care. According to Celia Deane-Drummond, religious ecologies seek to recover our sense of place on the earth, a reminder that the earth is our common home, and that the story of the earth and that of humans are one. This they do, unlike modern science, by uncovering the basis for our proper relationship between the origin of the earth (God), ourselves (humanity), and the environment (cosmos).[21] Grim and Tucker think, rightly in my view, that in a Christian mind-view of nature, the idea of ecology locates humans within the horizon of emergent, interdependent life and does not view humanity as the vanguard of evolution, nor as the exclusive fabricator of technology, nor as a species apart from nature.[22] Christian care for creation then requires a theology of inclusiveness and earthiness derived from a holistic worldview and theistic spirituality such as that of Africans.[23]

African Theocology: A Synthesis of African Christian Motivational and Scientific Ecology

Proposing African Christian theocology as a religious and scientific ecology that may motivate creation care is a response to the search for holistic approaches in combating the global ecocrisis from an African perspective.

17. Tucker and Grim, "Series Foreword," xx.
18. Howell, "African Spirituality and Christian Ministry," 1.
19. Agrawal and Gibson, "Enchantment and Disenchantment," 649.
20. Deane-Drummond, *Eco-Theology*, xi–xii.
21. Deane-Drummond, *Eco-Theology*, xi–xii.
22. Grim and Tucker, *Ecology and Religion*, 62.
23. Blasu, "'Compensated Reduction' as Motivating," 24.

Ben-Willie Kweku Golo observes that while the Western church responds to ecocrisis, particularly climate change in various forms, African Christianity is yet to engage actively with environmental activism[24] and the reconstruction of a creation care theology. M. L. Daneel was uncertain how, in view of Western Christian cosmology, African indigenous (primal and Christian) religious consciousness of the environment could contribute anything significant to the development of global environmental ethics. But he was certain that his publications would challenge and inspire someone "in the common quest of Earthkeepers worldwide to heal the Earth."[25] Perhaps I am one such challenged and inspired disciple of Daneel—a result of his published experiences in Africa. My interest in this study heightened in October 2013, when I learned that "religion provides strong motivation for human protection of the environment,"[26] with a clear case of Daneel's Zimbabwean Earthkeepers. The Zimbabwean earthkeeping projects were in response to their deforestation challenges, and were developed based on African indigenous religious cosmologies and ethical regulations—both primal and Christian.[27] Intrigued by their example, I intuitively described their work as "African theocology" and shared with some colleagues my hope to explore it further as an alternative to the environmental science course in our African universities, especially at Presbyterian University College, Ghana (PUCG).

By African theocology I envisage the study of the relationships between God (the Supreme Being) as Creator and his creations (human and nonhuman); particularly, the role of humanity in these relationships from the perspective of God, and conservative science and in the context of African religiosity. Independently and concurrently, Howard Kris Carter in Australia also defined theocology as "a mix of theology and ecology, the study of God and the study of the ecosystems."[28] As an academic discipline theocology is essentially to be an approach to doing both religious and scientific *ecology*. It is, therefore, essentially an *ecological* study, but recognizes that ecological or environmental science "needs to be in dialogue with other disciplines," especially African religions and religious ethics, "in seeking not only comprehensive solutions," but moral and missional impulsion to solving "both global and local environmental problems."[29] In other

24. Golo, "Redeemed from the Earth?," 352.
25. Daneel, *African Earthkeepers*, 106.
26. Daneel, "Zimbawe's Earthkeepers," 202–7.
27. The Christians in the projects belonged to sects of African Indigenous Christianity (AIC).
28. Carter, "Looking Up," para. 4.
29. "Forum on Religion and Ecology at Yale," para. 1.

words, I envision African theocology as a synthesis of Christian theistic and scientific ecologies; it is to be studied by exploring conservation science, African religious worldviews, ecoregulations, and rituals interpreted as much as possible with biblical texts in order to "broaden understanding of the complex nature of the current environmental concerns."[30] The ultimate hope of African theocology as a Christian curriculum is to encourage and provide Christian motivation to morally care for the earth as a mission of the church and church-based institutions. This involves the regaining of God's perspectives on ecological studies through, as much as possible, interpreting both the content of ecological science and African religious ecoknowledge with Christian Scriptures.

The adjective "African" places the conception of theocology in its own genre. "African" emphasizes the view that developing a Christian theology for creation care needs to be done "from within our cultural traditions."[31] Pope Francis emphasizes: "all of us can cooperate as instruments of God for the care of creation, each according to his or her own culture, experience, involvements and talents."[32] It is advantageous then for students of ecological studies in Africa to appreciate ecological phenomena and categories in their natural or indigenous African cultural contexts as a substratum and anterior to the modern scientific discourse. In their separate works, both Harry Agbanu and Patrick Curry suggest that previous cultural experience of valuing the natural environment is a necessary factor fundamental to appreciating scientific conceptions thereafter.[33] The Sokpoe-Eυe, like many other African primal cultures, hold a holistic view of the world, where community consists of the visible and invisible, of human and nonhuman members in diverse, but clearly known and ethically valued, structural relationships.[34] Bryant Myers observes, "while in no way the same, the worldview of the Bible is closer to the worldview of traditional cultures," such as in Africa, "than it is to the modern worldview"[35] of the West. They are both not only holistic, but also

---

30. "Forum on Religion and Ecology at Yale," para. 1.

31. Deane-Drummond, *Eco-Theology*, x.

32. Pope Francis, "Encyclical Letter *Laudato Si*," 13.

33. Harry L. K. Agbanu, in an interview with me at the University of Ghana, Legon, 23rd November 2015, opines "In fact I don't see how environmental science can be taught in a vacuum. It must be taught with people's understanding of the environment [and how it functions] from their indigenous cultural context. Then science can help them to improve on it." In addition, Curry (*Ecological Ethics*, 174) believes that global ecological ethics "must be encouraged and articulated on the basis of what is ecological and ethical in what people already value, know and do where they are."

34. Kalu, "Precarious Vision," 42.

35. Myers, *Walking with the Poor*, 8.

suggest strong bonds between humanity and the materiality of earth, particularly as demonstrated when Christ—"the creator, sustainer, and redeemer of creation . . . became flesh," or an earthling.[36] In reconstructing African theocology, I see three ecoethical concepts of the Sokpoe-Eʋe which have affinity with biblical worldviews and hence can support theologizing about Christian creation-care. These include their African theistic and precarious views of creation, their African sense of kinship with creation, and their African anthropocentric valuing of humanity.

African Theistic Cosmology as Preparation for Christian Theocentric Ecology

Although a significant amount of ecological degradation is anthropogenic,[37] humanity has the ability to change and work together in building our common home, the earth. But, as Pope Francis asserts, for humans to change and develop the right attitudes toward the care of creation, sustaining and renewing the earth, they need to be motivated.[38] A primary motivation for Christians to care for creation is inferable from the teaching that Christians are to obey and please God as well as seek his righteousness as their love for God in all their relationships in the ecosystem. This implies a theocentric worldview. Theocentrism is basically a Judeo-Christian ideology, connected to theocracy.[39] Richard A. Young rightly points out that a Christian theocentric worldview teaches that God is the center of the universe; the source and upholder of meaning, purpose, values, and ethics, as well as the unifying principle of the cosmos. Everything exists for the sake of God and to serve his purposes.[40] Christian theology affirms that there is only one God, whom the Christian must not only love with all the soul, heart, and might (Deut 6:4-6), but also look up to for help in all situations. God is the Creator, Owner, Sustainer, Protector, and Redeemer of the Earth and all there is in it (Ps 121:1-8).

As noted earlier, an ecological implication of theocentric affirmation is the Christian's urge to love God (John 14:24) by obeying and pleasing him in all things, including ecological responsibilities. The urge to love or fear God in all things becomes the motivation in theocentric cosmology for ecological actions, including ensuring ecological relations that reflect

36. Myers, *Walking with the Poor*, 8.
37. Curry, *Ecological Ethics*, 201–2.
38. Pope Francis, "Encyclical Letter *Laudato Si*," 12.
39. Victus, *Eco-Theology and the Scriptures*, 119.
40. Young, *Healing the Earth*, 128.

the righteousness of God in the ecocommunity (Matt 6:33). Richard Young argues that theocentrism encourages us to put faith and trust in God for the ultimate solution to our ecocrisis; it provides a reason for the existence of every creature; and it produces a holistic view of life, since everything is related by virtue of its being created by God.[41] Concerning African Christian motivation for reducing deforestation, for instance, I have averred elsewhere that the African theistic cultural self-understanding of life, interpreted by the gospel, contains a motivation that is anterior to the motivation of the sorts of economic gains the West promotes.[42] I stated further that a theistic, holistic, and precarious worldview can enable the African Christian to naturally appreciate the need for reducing deforestation from a religious awe of God, first and foremost, and hence, to act in obedience to seek first God's kingdom in all things (Matt 6:33). As I have argued, failure to let God's concern for his environment to be central in reducing deforestation could be classified as ecological sin.[43]

From their African primal theistic religious perspective, the Sokpoe-Eυe see "ecological sin" (*busu*) in terms of the breaking of ecological taboos, incurring the wrath of either *trɔ̃* (divinity/earthly deity) or *Mawu* (God), which results in *trɔ̃ku* (divinity-caused death). In separate interviews, both Enyi Avenorgbo[44] and Kofi Avinyo Atiglo[45] are very certain from their primal, religious self-understanding and belief in divinities that harvesting prohibited trees from sacred forests is not only an ecological sin, but inevitably incurs *trɔ̃ku*. Similarly, Sokpoe Christians like Winner Eworyi and Martin Doade will not enter a sacred forest let alone harvest firewood for fear of *trɔ̃ku*.[46] Thus, the fear of the spiritual entities, especially *trɔ̃* (earthly deity), is a significant impulsion for creation care among the Sokpoe-Eυe. The theistic spirituality and its related ecocultural self-understanding of the Sokpoe-Eυe, in my view, is a preparation for reconstructing a biblical theology in caring for creation by redefining the object of the fear that tends to be an ecoimpulsion. To reiterate, the Sokpoe-Eυe hold not only a theistic and holistic, but also a precarious view of the world.[47] The prevalence of various spiritual entities through all three dimensions of space (sky, the earth's

---

41. Young, *Healing the Earth*, 128.
42. Blasu, "'Compensated Reduction' as Motivating," 24.
43. Blasu, "'Compensated Reduction' as Motivating," 24.
44. Enyi Avenorgbo, interview at Sokpoe-Elavanyo, 2 February, 2016.
45. Kofi Avinyo Atiglo, interview at Sokpoe-Elavanyo, 15 February, 2016.
46. Winner Eworyi, interview at Sokpoe, 2 February 2018; Martin Doade, interview, Sokpoe, 2 February, 2018.
47. Kalu, "Precarious Vision," 42.

surface, and below the earth) of the *xexeme* (gecosphere) in their worldview underscores their conception of ecological ethics and praxis as not only religious, but specifically theistic.[48] Kalu describes it as "the sacralization of the environment."[49] John S. Mbiti's assertion that the daily life of the African people is notoriously religious, and that religion permeates all departments of life so that it is not easy or possible to isolate them,[50] reinforces that African worldviews, which underpin African religiosity as a cultural phenomenon,[51] are characteristically religious and theistic.

Yet in ordinary life experiences the primal African worldviews are not necessarily theocentric, particularly since not all cultures have God-dominant worldviews. But even in deity- or God-dominant religious systems, as among the Sokpoe-Eʋe, the Supreme Being, or God, is not necessarily central, pivotal, or the frontline power that people focus attention on for immediate and direct reach; the subordinate divinities and the ancestors may even be more paramount. Discussing environmental management from indigenous resources among the Kikiyu of Kenya, Julius Gathogo premised his argument that the indigenous Kikuyu people were encouraged to preserve the environment by their belief in the sacredness of nature: "Therefore, the ecological concern for people in Mutira . . . was tantamount to co-working *with* [not *because of* or *for*] God."[52] They practice an environmental management that is, at best, based on theistic religious, but not theocentric, cosmology.

Simon Victus opines, and it is practically a considerable view, that Christian theocentrism may be an ultimate solution to the present divergent views for tackling the global ecocrisis. It encompasses the concerns of the other views, without doing injustice to any of them and without acquiescing in their shortcomings.[53] I perceive that for the Sokpoe-Eʋe Christian, who holds a similar God-dominant worldview as the Bible, a paradigm shift from theistic religious ecopraxis toward theocentrism may not be impossible, though not without initial challenges. Reinterpreting the fear of deities with love for God the Supreme Being will influence our human responsibility to the environment and our expression of faith that the whole cosmos is a living revelation of God. I know very well that in Africa the *fear* of deities

48. Blasu, "Christian Higher Education as Holistic Mission," 134.

49. Ogbu Kalu, "Sacred Egg," 240.

50. Mbiti, *African Religions and Philosophy*, 1. See also Mbiti, *Introduction to African Religion*, 30.

51. Kalu, "Sacred Egg," 229.

52. Julius Gathogo, "Environmental Management and African Indigenous Resources," 1. Emphasis mine.

53. Victus, *Eco-Theology and the Scriptures*, 119.

is a stronger impulsion than *love* for God, and may challenge a paradigm shift to theocentric ecoethical praxis. Stella Xɔmeku, an elder of the Christ Evangelical Mission Church (CEM), in an interview, was very certain that, considering the quick action and, hence, the fear of deities in our worldview, a sudden change to the "fear of God in preserving sacred forests will not work; God does not punish quickly like *trɔ̃* (deity)."[54] Nevertheless, I believe that engaging the Bible with our theistic spirituality can produce a theocentric ecological theology for creation care since there is God/deity dominance in both the biblical and the Sokpoe-Eʋe worldviews. I call such theology African theocology, and a commitment to teaching and practicing African theocology as a curriculum for ecointegrity education in both the church and the Christian academy can be of great help.[55] The plausibility and possibility of this hope and expectation is further evidenced in the affinity between African and biblical concepts of kinship with nature.

African and Biblical Views of Human-Earth Kinship: Potential Motivation for Ecopraxis

Kinship with nature is a biblical truth presented in both the Old and New Testament stories or teachings about creation, its sustenance and hopeful redemption. Gillian Bediako claims it was "abandoned inadvertently due to erroneous and ignorant Western Christian theology, but recoverable in African primal spirituality."[56] An example of strong African primal religious belief in kinship with nature, which can be reconstructed for Christian theocology, is observable in the birthing rites of the Sokpoe-Eʋe. The ceremony involves burying the umbilical cord and placenta in the soil, placing baby nakedly on bare soil, and ensuring it urinates onto the soil. All these symbols are to inform and ground the baby as an earthling, which primes it for creation care as it grows into responsible status, learning from and practicing the ecoculture of its ecocommunity.

That both humankind and nonhuman creatures are kin is noticeable in the Christian views and teachings on creation and salvation. From creation stories it is obvious that both humans and nonhuman creatures are products of the same material substance, elements of earth. In the biblical creation story in Genesis 1:1—2:4a, soon after their emergence the

---

54. Stella Atsuƒe Xɔmeku, interview at Sokpoe, 29 February, 2016.

55. I have developed and proposed one such curriculum for the Presbyterian University College, Ghana in 2017. See Blasu, "Christian Higher Education as Holistic Mission," 188–89.

56. Bediako, "Primal Religion and Christian Faith," 14.

primordial earth and its waters were enlisted by God to participate in his ongoing creative activity. The earth responded to God's call and put forth vegetation and various kinds of beasts (Gen 1:11, 24), and the waters and air (atmosphere) produced swarms of aquatic living creatures and birds, respectively (Gen 1:20).[57] Then in the story in Genesis 2:4b–25, God formed humanity (*adam*) of the dust from the ground (*adamah*; Gen 2:7a). Thus, all are created by the same God, from the same materials, and by virtue of divinely endowed function they all serve in mutual or symbiotic vocation to sustain all life on earth.

In Ghana, concerning forests, for example, it is commonly said: "when the last tree dies the last man dies." To emphasize the fact that the only possible breach of this mutual function is anthropogenic, American Indians of the Cree tribe formulated this statement more fittingly: "Only when the last tree has been cut, the last river is poisoned, and the last fish is caught, one will realize that one cannot eat money."[58] I have argued elsewhere that I do not deny the place of monetary motivation in reducing deforestation, for African religious motivation is anterior to, and more holistic than, monetary enticements.[59]

As in creation, so also is the doctrine of salvation. Humanity's kinship with nature is so apparent in Africa that we need to shift from a personal to a cosmic view of salvation. From a theocological point of view, Kweku Golo decries how traditionally Christian salvation theology emphasizes only individual encounters with Jesus Christ leading to repentance and redemption from sin, the transformation into a new person, and empowerment through the Holy Spirit to live a new life. Salvation, in that sense, "is understood spiritually and heavenwards and pictured as an eschatological blissful state in which Christians participate."[60] Yet ecologically this means removing humanity from creation, regarded as corrupted by sin from its goodness, into the "transformed and sanctified state through the sacrifice of Jesus Christ."[61] The result is that the cosmic shape of salvation is minimized or even entirely overlooked in Western Christianity. Golo continues that the otherworldliness concept of Christian salvation theology spilled over to Africa through Western missionaries and not only found a home, but was even radicalized due to the holistic but precarious African cosmology.[62] In

---

57. Schwarz, *Creation*, 170.
58. Nehring, *Ecology*, 254. See also Victus, *Ecotheology and the Scriptures*, 6.
59. Blasu, "'Compensated Reduction' as Motivating," 24.
60. Golo, "Redeemed from the Earth?" 356.
61. Golo, "Redeemed from the Earth?" 356.
62. Golo, "Redeemed from the Earth?" 356.

this cosmology, multifarious spirits vivify the cosmos, but malevolent ones may discourage belief in Christ and living a fulfilled Christian life. Hence, salvation is understood as redemption from sin and evil forces, transformation into a new person, and turning away from creation by waging war against the evil forces within it.[63]

Birgit Meyer observes among the Peki-Eʋe in Ghana that the pietistic salvation message of the Norddeutsche Missionsgesellschaft (NMG) did not only "demonise" the "Eʋe gods and spirits," but also drew a "boundary between Christianity and Eʋe religion,"[64] and, hence, ecocultural self-understanding. It is not surprising, for instance, that despite the strong kinship belief exhibited in primal religious birthing rites at the Sokpoe ecological area in Ghana, the Christians there, also nurtured mostly with western pietistic Protestant theology by the Basel missionaries (BM), expunged the symbolic grounding rituals from the outdooring liturgy, because they are *egodotɔwo tɔ* (for non-Christians and evil).[65]

Furthermore, the mercantilist manner of presenting the salvation message in Africa resulted in an anthropocentric utilitarian notion of salvation that has reduced the earth to commoditization, according to Golo. The earth, then, has no spiritual significance, save for being an abode of demonic spirits which must be overcome if humans would live in material prosperity—a sign of salvation found especially in neo-Pentecostalism.[66] In her analysis of the Peki-Eʋe's reasons for converting to Christianity, Birgit Meyer argues that the "Christian religion was attractive because it offered the material means to achieve a prosperous and relatively high position in colonial society."[67] Meyer concludes that "demonisation by no means implies that the former gods and spirits [in Eʋe cosmology] will disappear out of people's lives," including Christians.[68]

In a nutshell, the missionary dualistic worldview and otherworldly salvation theology in contemporary Africa has not succeeded in eroding how the primal African conceives of creation.[69] Neither did it succeed in completely unravelling the African Christian's holistic but precarious cosmology. Thus, the Eʋe Christian is left in a religious dilemma of how Christian salvation theology pragmatically reconciles monotheism with Africans' ontological

---

63. Golo, "Redeemed from the Earth?" 351.
64. Meyer, *Translating the Devil*, xvii.
65. James Cudjoe, Church of Pentecost, interview at Sokpoe, 26 June 2016.
66. Golo, "Redeemed from the Earth?" 356.
67. Meyer, *Translating the Devil*, 11.
68. Meyer, *Translating the Devil*, xvii.
69. Golo, "Redeemed from the Earth?" 352.

experiences of other spiritual forces vivifying the gecosphere. Elsewhere, I described African Christians who are unable to appropriately engage the fear of primal religious deities with christological faith, especially among the Sokpoe-Eʋe, as the "Culturally Passive or Suspended," people who are undecided for or against primal religious ecoethics and praxis.[70]

Jesus' death and resurrection brings humans into a state of reconciliation with God and also opens the way for the reconciliation of all things to God. The earth does not only suffer God's curse on, and destruction of, humanity due to the latter's disobedience or sinfulness (Gen 3:17–19; 6–8), but also co-enjoys God's promise of no destructive deluge again (Gen 9:7–17) and hopeful redemption (Rom 8:19–23). Saint Paul teaches that creation waits in eager expectation for the children of God to be revealed. For creation was subjected to frustration, not by its own choice, but in hope that it will be liberated from its bondage to decay and brought into the glorious kingdom of the children of God by the same redeemer, who had permitted the subjection. So, the redemption of the earth for eternal existence is also linked with that of humanity—to have full life (John 10:10). Before the consummation of redemption, both humanity and creation groan from the consequences of their sin,[71] which threaten the destruction of their life, painfully, to decay. However, to the Christian, biogeophysical cyclic processes that ensure the equilibrium of the gecosphere for life are divine provisions for healing.

Christian doctrine asserts that God the Father heals his people and the land (2 Chr 7:14–15), and that Christ, as the perfect human (the Son of God), identifies with the whole of creation (Col 1:15–20). Strengthened by the Holy Spirit (Luke 22:43), Jesus atoned for the sins of creation with his blood on the cross (Isa 53:4–5). Ernest Lucas believes that Jesus' healing miracles should not be seen from a purely human-centered perspective. They are signs of the coming renewal of the whole created order.[72] Thus, despite threatening devastations of terrestrial life as consequences of sin, all creation hopes for the triune God's active sustenance and working toward consummation of redemption for eternal fullness of life. In this way, Christ, the second Adam, is "keeper" of the earth more perfectly than the first Adam, since in him "all things hold together" (Col 1:17).[73] That Christ holds all creation together plausibly interprets the African concept of human kinship with nature to underscore and urge Christian care for creation as people in the image of the

---

70. Blasu, "Christian Higher Education as Holistic Mission," 160.

71 As we discussed earlier, the sin of humanity was imputed for creation also and both started suffering its consequences of varied "painful" experiences since the time of Adam (Gen 3:16–19).

72. Lucas, "New Testament Teaching," 94.

73. Daneel, *African Earthkeepers*, 215.

second Adam (Jesus Christ) and not the first Adam. The first Adam (fallen humanity) became ecologically bankrupt due to misinterpretation of its cultural self-understanding as the image of God.

African Anthropocentric and Biblical Theomorphic Valuing Can Impel Creation Care

One source for studying African religiosity and philosophical thought, such as ecological valuing, is by analyzing maxims, including indigenous names. In the study for this paper I reflected on both the literal meaning and axiological implication of the name *Amewuga*—"human is more valuable than money." They mean that in the ecosystem only humans have intrinsic values in their views. The reasons include observations that humans possess qualities of emotionality, rationality, lingual skills, initiative taking, and relatively higher versatility, which nonliving aspects of nonhuman creation lack. Further evidence of ecovaluing was given about the traditional order of rescuing some creatures as flood victims: human, goat, tree, and gold nuggets. In the interview, Margaret Asuma, a deaconess of the Church of Pentecost (COP), was among 61 percent (fourteen out of twenty-three interviewees) who followed the descending order of human, gold, and goat before the tree in rescuing flood victims.[74] She gave her reasons as follows:

> Although God created all things before humanity, yet human beings are more important among all the things. Gold/money follows because money enables humans to be complete; the goat will be for food, and so will be the third. As for the tree it may drown. No problem.[75]

For Deaconess Margaret Asuma, like those with similar views in the study, the human being has life and is created in God's image; both the rescued human and the rescuer in the scenario are fellows, *Homo sapiens*.[76] The analyzed responses indicate that like the primal religionists, the focus group participants' ecovaluing of creation mixes biocentrism with zoocentrism, but is more practically anthropocentric. In other words, the Sokpoe-Eʋe Christian values and practices ecological ethics ultimately with an anthropocentric

---

74. Actually 5 of 23 (22 percent) selected "goat" instead of "tree" to be last in the rescue order, but the reasons advanced were almost like those (9 of 23, 39 percent) who would rescue a tree last. In both groups the emphasis is on anthropocentrism and biocentrism.

75. Margaret Asuma, interview at Sokpoe-Elavanyo, 25 February 2016. See appendix A17 for transcript.

76. Mary Adzabeng, interview at Sokpoe, 21 June 2016.

view of the ecosystem. Agbanu's study on the indigenous environmental ethics of the Mafi-Eʋe of North Tɔŋu in Ghana also makes similar observations. He concludes that their ecological ethics is anthropocentric, but mixed with biocentrism and ecocentrism.[77] Ogbu Kalu seems to express surprise at this, wondering that "In spite of the remarkable awareness of spiritual forces, the African places man (sic) at the centre of the universe."[78] And Kwame Bediako makes a similar observation for most African cultures.[79] My proposition is that, in a sense, African anthropocentrism has connotations of biblical theomorphism, which can be engaged with the gospel to develop a Christian theology of "vicegerency" for creation care.

Two Christian teachings that are controversial and yet truly underscore and mandate Christians to practice creation care are the theomorphic and steward ascriptions to humanity. Biblical texts such as Genesis 1:26–31 and Psalm 8:4–8 describe humanity as a being in the image of God or after his likeness (*imago Dei*) or a little lower than God and given dominion as stewards over God's handiworks. But ecologically these passages have been problematic. Often, they are interpretively reduced to negative inferences of the *imago Dei* and its associated dominion theses, carrying the same ecological implications as African anthropocentrism. The dominion thesis has been the cause of many debates, accusations, and labelling of Christianity as an anti-ecological religion and great contributor to the current ecocrisis.[80] However, it is arguable that the problematic nature of this doctrine in itself suggests its potency also as an impulsion for Christian ecocare praxis. Implicit in it is also an irrefutable divine truth: humanity has theomorphic qualities; we are created in the image of God (as *imago Dei*). Thus, as Schwarz suggests, it is important that we discern what it means to be created in God's image,[81] particularly in reconstructing Christian theocology with regard to the place or role of humanity in creation.

According to Solomon Victus, in the Old Testament, two Hebrew words—*tselem* (image) and *demut* (likeness)—are used for the expression "created as *imago Dei* (God's image)." The former refers to a physical form like an image or idol, which cannot be conceived in Hebrew thought, given their strict adherence to the Decalogue; and the latter is simply a similitude. Therefore, Victus argues that in the priestly text of Genesis these words are

77. Agbanu, "Environmental Ethics in Mafi-Eʋe Indigenous Culture, " iii.

78. Kalu, "Precarious Vision," 41.

79. Bediako, *Jesus in Africa*, 92.

80. White, "Historical Roots of Our Ecological Crisis," 1203–7 is the most noted accuser of Christianity's ecological heritage based on this passage. See also Curry, *Ecological Ethics*, 33.

81. Schwarz, *Creation*, 179–80.

just a "synonymous parallelistic statement," being a literary genre of Hebrew poetry though some try to differentiate between them.[82] In any case, since in the biblical creation stories only humanity was created in and designated as *imago Dei*, it is not difficult to conclude that humans "are very close to God"[83] and have an elevated position (dominion) over against other creatures.[84] For instance, humans, by virtue of their genetic programming and neurobiological possibilities, have been endowed with unique capabilities with which they shape the earth for good or bad, and hence are co-creators with God; but also, unlike God, they are destroyers of creation. As John Chryssavgis argues, "the human person is characterized by paradoxical dualities . . . created yet creative."[85] But this is not to be understood as deifying humans or making them domestic servants of the gods as in Mesopotamian and Babylonian mythologies, respectively,[86] despite such a suggestion in Psalm 82:6a: "You are gods, sons of the Most High, all of you."

Leonardo Boff argues from an Eastern Orthodox Christian understanding that when viewed in God's perspective and light, all things are sacraments, and "the word, human beings, and things are signs and symbols of the transcendent."[87] Boff implies, agreeably, "that the meaning of *imago Dei* and humanity being given dominion need to be viewed and interpreted in the perspective and light of God in reconstructing Christian theocology, for instance. Doing so brings out that both humans and nonhuman creation (things) are both matter with same moral and sacramental value; both then matter in the sight of their maker, God."[88] In other words, as Victus concludes, human needs and desires are not to be considered in any sense higher than those of the rest of nature (and I add, without reference to God), for the human is nothing more than a part of nature before God.[89] Thus, even if humans are given the special status of the image of God, we do not have the mandate to be arrogant, dominant, and oppressive.[90] Rather, the *imago Dei* means to be God's vicegerent, "to act [graciously] in God's place, as his administrator and representative" of the gecosphere; and, from the New

---

82. Victus, *Eco-Theology and the Scriptures*, 83.

83. Victus, *Eco-Theology and the Scriptures*, 83.

84. Schwarz, *Creation*, 181.

85. Chryssavgis, "World of the Icon," 87.

86. Schwarz, *Creation*, 181.

87. Boff, *Sacraments of Life-Life of Sacraments*, 49-51. See also Chryssavgis, "World of the Icon," 90.

88. Boff, *Sacraments of Life-Life of Sacraments*, 90.

89. Victus, *Eco-Theology and the Scriptures*, 103.

90. Victus, *Eco-Theology and the Scriptures*, 83.

Testament, it is "to be ethically shaped in conformity with God and to act in a manner for which God serves as the prototype [ecologist] (Phil. 2:5; Rom 15:5)."[91] For Schwarz, being in God's image and having dominion does not imply a special ontological quality, but an assertion about the ecological function of humanity. It is a role that contains authority, but at the same time humility to represent and model God's loving and caring nature, which then makes us truly creatures and also co-creators.[92] That humanity's dominion authority demands humility and moral accountability is because it is derived; only God wields actual dominion over creation as Peter succinctly spells out: "To God be dominion forever and ever" (1 Pet 5:11).

To be created in God's image is also a role with moral responsibility. The first Adam was thus given functional instruction to "work" (use) and yet "keep" (care for or sustain) the land (Gen 2:15); but he cannot have full freedom to do whatever he wants to do, being limited and forbidden to eat of the fruit of the tree of the knowledge of good and evil (Gen 2:17). So, Adam can function only within the limits of and according to God's ecological ethics such as sustainable use of resources (Gen 2:15) and Sabbath rest (Gen 2:2). His authority is derived from God and is "not a licence to anthropocentrically exploit creation and subjugate it to his desires."[93] Norman L. Geisler discusses the materialistic, pantheistic, and Christian worldviews in his chapter on ecology and he concludes that in Christianity God is the creator and humans are the keeper of God's glorious world. It is our duty to keep and not corrupt, to preserve and not pollute.[94] Doing otherwise is ecological sin.[95] As I see it, the fate which befell the first *Adam* (the fallen humanity) has successfully worked in all to fail woefully in keeping the land. It would be left for Christ, the second *Adam*, as Daneel says, to help all to be a keeper of the earth more perfectly than the first Adam since in him "all things hold together" (Col 1:17).[96]

The theocological implication is that just as the first Adam represents fallen humanity so Christ, the second Adam, is a figure of the regenerated humanity (Rom 5:18–19), the head of the new creation (2 Cor 5:17). Jesus, unlike sinful Adam, is the exact imprint of God's nature (Heb 1:3). Jesus then demonstrates the true picture of being in God's image, vicegerent,

91. Schwarz, *Creation*, 182.
92. Schwarz, *Creation*, 183.
93. Schwarz, *Creation*, 182.
94. Geisler, *Christian Ethics*, 333.
95. Metropolitan John (Zizioulas) of Pergamon, "Pope Francis's Encyclical *Laudato Si'*: A Comment."
96. Daneel, *African Earthkeepers*, 215.

and, therefore, the one with the perfect creation care disposition. Jesus the Word as the creator is also the Word become flesh (John 1:14).[97] In the flesh, Jesus is the one who truly reveals God, "the image of the invisible God, the firstborn of all creation" (Col 1:15).[98] He then is the one in whom the image of God (*imago Dei*) is restored, and consequently he makes believers "a new creation" (2 Cor 5:17).[99] What it means is that as in Christ we are being reconstituted with the true *imago Dei* nature, and it is imperative that we are motivated as vicegerents to care for creation in Christ's way. We are created in Christ Jesus for good works, beginning with creation care (Gen 2:15), which God prepared beforehand so that we should walk in them (Eph. 2:10b).

In practice, humanity must care for creation because humans are immersed in nature, inexorably woven into profound relationships that constitute life on Earth.[100] Humans are called to a priestly role of responsibility and vicegerency toward creation. This sense of a call to responsible earth-care made the Ecumenical Patriarch Bartholomew speak of human degradation of the environment as ecological sin—disobedience to our ecological call.[101] We sin against God, who created to reveal himself and provide a good *oikos* (home) for creaturely life; we sin against the environment and ourselves by blocking the environment from serving its divine purpose of nurturing us.[102] Only humans in all creation will be held accountable for the way(s) in which we have exercised our responsibility to and for creation.[103] However, in obeying the call to care for creation we must recognize creation's limited carrying capacities and the limits of our caring,[104] by adhering strictly to God's ecovalues and ecoethics. Conventionally, a discourse on engaging God's ecovalues and ethics in creation care is often described as ecotheology. Why then African theocology?

---

97. Keitzar, "Creation and Restoration," 59.
98. Keitzar, "Creation and Restoration," 60.
99. Keitzar, "Creation and Restoration," 61.
100. Bauckham, "New Testament Teaching," 100.
101. Allison, Review of *Ecology and Religion*.
102. Geisler, *Christian Ethics*, 333.
103. Berry, "Christian Approach to the Environment," 73.
104. Curry, *Ecological Ethics*, 35.

African Theocology and Ecotheology: A Comparison of Nomenclature and Implications

Admittedly, my idea of theocology as an evolving, twenty-first century study is similar in many respects to that of ecotheology, the conventional way of naming the Christian religious approach to studying ecology. Both are comparative in terms of their ultimate purposes and major thematic contents, because they are conceived as subsets or corollaries of the theology of nature. Like African theocology, ecotheology is an emerging discipline of study[105] which originated in the second half of the twentieth century,[106] but which became more pronounced in the 1990s.[107] My motivation to move from ecotheology to theocology is to let African theocology be perceptible, understood, and practiced also as an *ecological science*, but starting with grounding in God as the Creator of ecosystems. It is a contextual motivation arising from an experience in lecturing and analyzing concerns of some environmental science students and lecturers that has led to writing this paper. The first concern is their dilemma of conceiving and locating ecotheology, a subject they see as being theological rather than the science course they paid and registered for, to be an alternative to environmental science. Specifically, some teachers and nontheological students argued against the integration of Christian religious thought in the study of environmental science, a mandatory course aimed to promote Christian moral environmentalism at Presbyterian University College, Ghana (PUCG). Students saw the integration of biblical moral thought in environmental science as turning a supposedly pure science subject into a theological study, to their chagrin. The second concern was from some lecturers who feared that the environmental science course would lose accreditation as a purely scientific study and/or be considered illegal if a theological motif was encouraged for the purpose of effectuating religious moral environmentalism. While appreciating the role and need of Christian theology in moral environmentalism, yet to avoid illegalities and loss of accreditation as a science course, respondents suggested a renaming and a reaccreditation of environmental science to reflect the moral objective with a theological motif.

Ecotheology readily came to mind, but proposing it as an alternative to the environmental science was not helpful in the situation. As alluded to earlier, it really appears and sounds on the surface like more of a theological reflection than an ecological science subject and belongs to

---

105. Grim and Tucker, *Ecology and Religion*, 63.
106. Grim and Tucker, *Ecology and Religion*, 87.
107. Cunningham and Saigo, *Environmental Science*, xv.

the academic field of theology more than ecological science. Solomon Victus states that ecotheology is "fundamentally a contextual *theology*."[108] It emerged in response to a twentieth-century (1960s) ecocrisis which was created by the free market economy, neoliberal economic policies of states and corporate sectors, particularly in the West, and which resultantly commoditized nature for profit making.[109] In this respect, "ecotheology speaks about a *theology*,"[110] and is hence studied as "a form of constructive theology"[111] rather than as an *ecological science*, though "it speaks from the perspectives of creation, ecology and human responsibility."[112] When the Protestant theologian Joseph Sittler first publicized ecotheology as an academic discipline, he considered it as a theological construct and so called it "a new theology of grace that included rather than excluded nature."[113] In other words, "ecotheology describes theological discourse that highlights the whole 'household' of God's creation, especially the world of nature, as an interrelated system (*eco* is from the Greek word for "household," which is *oikos*)."[114] Ecotheology is studied with various ideological positions or facets, including ethical positions such as anthropocentrism, biocentrism, ecocentrism, and theocentrism, and approaches such as process theology, ecofeminism, tribal theology, and animal theology.[115] In all cases, the emphasis of the approach is on *theology*. Therefore, the picture will be different for nontheology students and lecturers in an ecological science class by inversing the order of the name from eco*theology*, a theological course, to theo*ecology*, an ecological science course. The latter is designed to be delivered as an ecological science, but with emphasis on religious worldviews for the sake of providing a holistic comprehension and stimulation for moral and missional ecopraxis. In other words, theoecology is a study to explore "divine-human-earth relations,"[116] it is not earth-human-divine relations as seems implied by ecotheology.

Furthermore, ecotheologists claim that it is conceptually *not* natural theology (knowing God from nature); rather "theology of nature" (knowing

---

108. Victus, *Eco-Theology and the Scriptures*, 10 (emphasis mine).
109. Victus, *Eco-Theology and the Scriptures*, 10.
110. Victus, *Eco-Theology and the Scriptures*, 10. Italics mine.
111. "What is ecotheology?," https://www.youtube.com/watch?v=o4-KyANo3Kg.
112. Victus, *Eco-Theology and the Scriptures*, 10.
113. "Ecotheology," in *Encyclopedia of Science and Religion*.
114. "Ecotheology," in *Encyclopedia of Science and Religion*.
115. Victus, *Eco-Theology and the Scriptures*, xvii.
116. Tucker and Grim, "Series Foreword," xxiv.

nature from God's perspective).[117] Yet the prefix beginning it is eco-, which implies "our natural home," and tends to render ecotheology, at least from its nomenclature, similar to "natural theology." In that case the name contradicts the purpose, content, and the conception of beginning ecotheological study from God's perspective, as Creator, and on whose terms creation can be cared for. By inversing the word to theoecology so that the name begins with the prefix *theo-* and ends with the suffix *ecology*, the name more readily connotes the direction of study as from God's perspective on ecology. It is designed as a Christian missional and transformative curriculum, and hence, is based on *missio Dei*, the mission of God in and to all creation. But "the mission of God starts with God."[118] Thus by inversing the name I hope to emphasize the curriculum as still being "environmental science," or specifically "ecological science," except that a theocentric worldview emphatically underpins it and, as much as possible, is interpreted with Scripture or from Christian theological perspectives. African theocology is not simply a study of God from nature (though this is implicit), but also of how God relates with creation. On the contrary, it is an ecological science, a holistic study of our natural home, the earth, and our moral and missional responsibility to it from God's perspective as Creator, and as understood by engaging the gospel with African worldviews and ecoculture.

Conclusion

I have argued in this paper that implicit in the Bible and caring for the land is a call to develop a Bible-based theology that can undergird and motivate Christians toward ecological actions to sustain the integrity of creation. For African Christianity I propose that such theology needs to be an engagement of the African theistic religious worldviews with Christian Scriptures to produce African theocologies. I argued for the role of religion generally in creation care and also what I mean particularly by African theocology—studying about God and God's relations with his creation from God's perspective in the African context, to guide human moral responsibility to and for the environment. From my study among the Sokpoe-Eʋe of Ghana I deduced three elements in African ecological thought and praxis, which have bearings on the Bible and so may be preparations for reconstructing African Christian theocology. There can be, for instance, a reconstruction of African precarious but theistic and holistic views of the cosmos to provide theocentric impulsion for creation care, and African and biblical concepts

---

117. Victus, *Eco-Theology and the Scriptures*, 8.
118. Wright, "Holistic Mission," para. 13.

of kinship with nature which are potent for urging Christians to respect and care for creation. Similarly, the axiological closeness of both African anthropocentric and biblical theomorphic valuing of humanity can be engaged with the gospel for a theology of vicegerency. African theocology then hopes to provide a theocentric Christian impulsion for creation care. Despite recognizing cultural challenges in a paradigm shift to theocentrism, I believe that our commitment to teaching and practicing an African theocological curriculum at all levels of education in the church and Christian educational institutions is a good start.

Bibliography

"A Comment." http://www.goarch.org/news/metropolitanjohnpergamon-06182015.

Agrawal, Arun, and C. C. Gibson. "Enchantment and Disenchantment: The Role of Community in Natural Resource Conservation." *World Development* 27.4 (1999) 629–49.

Allison, Elizabeth. Review of *Ecology and Religion*, by John Grim and Mary Evelyn Tucker. *Society and Natural Resources* 29.6 (2016) 755–57.

Bauckham, Richard. "The New Testament Teaching on the Environment: A Response to Ernest Lucas." *Transformation* 16.3 (July 1999) 99–101.

Bediako, Gillian M. "Primal Religion and Christian Faith: Antagonists or Soul-mates?" *Journal of African Christian Thought* 3.1 (June 2000) 12–16.

Bediako, Kwame, "Scripture as the Hermeneutics for Culture and Tradition." *Journal of African Christian Thought* 4.1 (June 2001) 2–11.

———. *Jesus in Africa: The Christian Gospel in African History and Experience.* Cumbria, UK: Editions Clé and Regnum Africa, 2004.

Berry, Sam. "A Christian Approach to the Environment." *Transformation* 16.3 (July 1999) 73–74.

Blasu, Ebenezer Yaw. "'Compensated Reduction' as Motivating for Reducing Deforestation: An African Christian Theological Response." *Journal of African Christian Thought* 18.1
(June 2015) 18–27.

———. "Christian Higher Education as Holistic Mission and Moral Transformation: An Assessment of Studying Environmental Science at the Presbyterian University College, Ghana and the Ecological Thought of the Sokpoe-Eve for the Development of an African Theocology Curriculum." PhD diss., Akrofi-Christaller Institute of Theology, Mission and Culture, Akropong-Akuapem, 2017.

Boff, Leonardo, *Sacraments of Life, Life of Sacraments*. Washington, DC: Pastoral, 1996. Hessel, Dieter T., and Rosemary Radford Ruether, eds. *Christianity and Ecology.* Cambridge, MA: Harvard University Press, 2000.

Carter, Howard. "Looking Up . . . The Majesty of God and Our Place in Creation Psalm 8 Theocology; he Breath Print of God and Renewal of Creation Care in the Nature Psalms (Part 1)." Howard-carter.blogspot.com/2013/10/looking-up-majesty-of-god-and-our-place.html.

Chryssavgis, John. "The World of the Icon and Creation: An Orthodox Perspective on Ecology and Pneumatology." In *Christianity and Ecology*, edited by Dieter

T. Hessel and Rosemary Radford Ruether, 83–96. Cambridge, MA: Harvard University Press, 2000.

Cunningham, William P., and Barbara Woodworth Saigo. *Environmental Science: A Global Concern*. 6th ed. New York: McGraw-Hill, 2001.

Curry, Patrick. *Ecological Ethics: An Introduction*. 2nd ed. Cambridge, UK: Polity, 2011.

Daneel, M. L. *African Earthkeeper: Wholistic Interfaith Mission*. Maryknoll: Orbis, 2001.

———. "Zimbawe's Earthkeepers: When Green Warriors Enter the Valley of Shadows." In *Nature, Science and Religion*, edited by Catherine M. Tucker, 191–212. Santa Fe, NM: SAR, 2012.

Deane-Drummond, Celia. *Eco-Theology*. London: Darton, Longman and Todd, 2008.

Bergant, Dianne. *The Earth is the Lord's: The Bible, Ecology and Worship*. Collegeville, MN: Liturgical, 1998.

Eckersley, Robyn. "Beyond Racism." *Environmental Values* 7 (1998) 165–82.

"Ecotheology." *Encyclopedia of Science and Religion*. http://www.encyclopedia.com/education/encyclopedias-almanacs-transcripts-and-maps/ecotheology.

"Forum on Religion and Ecology at Yale." http://fore.yale.edu/files/Forum_Overview.pdf.

Francis, Pope. "Encyclical Letter *Laudato Si'* of The Holy Father Francis on Care for Our Common Home," http://w2.vatican.va/content/dam/francesco/pdf/encyclicals/documents/papafrancesco_20150524_enciclica-laudatosi_en.pdf.

Gathogo, Julius. "Environmental Management and African Indigenous Resources: Echoes from Mutira Mission, Kenya (1912-2012)." http://uir.unIsaac.za/bitstream/handle/10500/13127/Gathogo.pdf?sequence=1.

Geisler, Norman L. *Christian Ethics: Contemporary Issues and Options*. 2nd ed. Grand Rapids: Baker, 2010.

Golo, Ben-Willie Kweku, "Redeemed from the Earth? Environmental Change and Salvation Theology in African Christianity." *Scriptura* 111.3 (2012) 348–61.

Grim, John A., and Mary E. Tucker. *Ecology and Religion*, Washington, DC: Island, 2014.

His Holiness Bartholomew I, through John Zizioulas of Pergamon, on 'Pope Francis' Encyclical Letter *Laudato Si'* of The Holy Father Francis on Care for Our Common Home, on 24th May 2015: A Comment,', http://w2.vatican.va/content/dam/francesco/pdf/encyclicals/documents/papa-francesco_20150524_enciclica-laudatosi_en.pdf.

Howell, Allison. "African Spirituality and Christian Ministry: 'Discerning the Signs of the Times' in Our Environment and Community." In the 9th Kwame Bediako Memorial Lecture, at British Council, Accra, Ghana, 7th June 2017.

Kalu, Ogbu U. "Precarious Vision: The African's Perception of His World." In *Readings in African Humanities: African Cultural Development*, edited by Ogbu U. Kalu, 37–44. Ibadan, Nigeria: Fourth Dimension, 1980.

———. "The Sacred Egg: Worldview, Ecology, and Development in West Africa." In *Indigenous Traditions and Ecology*, edited by John A. Grim, 225–48. Cambridge, MA: Harvard University Press, 2001.

Keitzar, Renthy. "Creation and Restoration: Three Bible Reflections," In *Ecotheology: Voices from the South and North*, edited by David G. Hallman, 52–64. Eugene, OR: WCC, 1994.

Lucas, Ernest. "The New Testament Teaching on the Environment." *Transformation* 16.3 (July 1999) 93–99.

Mbiti, John S. *African Religions and Philosophy*. 2nd ed. Oxford: Heinemann, 1990.

———. *Introduction to African Religion*. 2nd ed. Long Grove, IL: Waveland, 1975.

Meyer, Birgit. *Translating the Devil: Religion and Modernity among the Ewe in Ghana*, Edinburgh: Edinburgh University Press, 1999.

Myers Bryant L. *Walking With the Poor: Principles and Practices of Transformational Development*. Rev. ed. Maryknoll, NY: Orbis, 2011.

Nehring, Andreas, ed. *Ecology: A Theological Response*. Madras: Gurukul Lutheran Theological College & Research Institute, 1993.

Özdemir, Ibrahim. "Toward an Understanding of Environmental Ethics from Qur'anic Perspective." In *Islam and Ecology*, edited by Richard C. Foltz et al., 45–58. Cambridge, MA: Harvard University Press, 2003.

Ptance, Ghillean. "Christians in Conservation," a press release for A Rocha, 30 March 2005. https://epdf.pub/between-heaven-and-earth-christian-perspectives-on-environmental-protection.html.

Schwarz, Hans. *Creation*. Cambridge: Eerdmans, 2002.

Son, Bong-Ho. "The Relevance of a Christian Approach." A paper presented at Christian Education in the African Context: Proceedings of the African Regional Conference of the International Association for the Promotion of Christian Higher Education, Harare 4-9 March. https://library.aiu.ac.ke/cgi-bin/koha/opac-detail.pl?biblionumber=35716. Accessed June 23, 2019

Tucker, Evelyn, and John A. Grim. "Series Foreword." In *Islam and Ecology*, edited by Richard C. Foltz et al., 96–105. Cambridge, MA: Harvard University Press, 2003.

Victus, Solomon. *Eco-Theology and the Scriptures: Revisit of Christian Responses*. New Delhi: Christian World, 2014.

"What is Ecotheology?" https://www.youtube.com/watch?v=04-KyANo3Kg.

White, Lynn, Jr. "Historical Roots of Our Ecological Crisis." *Science* 155 (10 March 1967) 1203–7.

Wright, Chris J. "Holistic Mission: What is Holistic Mission and Where Do We Start?" Lecture delivered to Wycliffe Bible Translators on 7th May, 2012, in Quezon City, Philippines.

Young, Richard A. *Healing the Earth: A Theocentric Perspective on Environmental Problems and Their Solutions*. Nashville: Broadman & Holman, 1994.

Part II

Understanding the Idea of ἔρήμος in Luke's Gospel: Implications for the Care of the Land in Ghana

Daniel Nii Aboagye Aryeh

Perez University College, Winneba

Abstract

The architecture of the concept of ἔρήμος is a theme that has its foundation in ancient Semitic religiocultural environments, and in the Old Testament. "Ερήμος" was conceived as a desolate place against human survival and at the same time a location for the presence of YHWH. In the New Testament, particularly the Gospel of Luke, the term "ἔρήμος" was largely engaged by the author to describe a divine setting for theophanies. This essay argues that Luke used "ἔρήμος" as a location where theophanies are experienced through prayer and ascetic life. This view is similar to the use of mountains for prayer by Pentecostal and charismatic church leaders, and the use of sacred places by African Traditional Religion. Hence, there is the need to develop a theology of ἔρήμος leading to the care of ἔρήμος in Ghana. The overall objective of this paper is to appreciate and contribute to scholarly discussion of the Lukan idea of "ἔρήμος" and promote the care of sacred places in Ghana.

Introduction

THE CONCEPT OF ἔρήμος (lit. "*herēmos*"), or "wilderness," is a common phenomenon in the New Testament, particularly the Gospel of Luke. The

Hebrew terms *midbār*, "desert which also mean steppe, grassland, e.g., Gen 37:22,"[1] and *ᵃrābāh* in a nonliteral sense (Isa 40:3) are also translated as ἔρημος in the Septuagint (LXX). According to Robert W. Funk, the non-literal use of "ἔρημος" is an ancient Near East (ANE) mythology, which partially found its way into the Bible.[2] The term occurs about 345 times in the Bible (LXX; NT), out of which a third is found in the Pentateuch, particularly Exodus and Deuteronomy,[3] due to Israel's activities with YHWH in the "ἔρημος," and Israel's readiness to leave the "ἔρημος" and begin to possess the promised land. In the Gospel of Luke, "ἔρημος" occurs ten times and it is translated "wilderness" or "desert" by the New Revised Standard Version (NRSV).

Ἔρημος is a synonym for "desert" and "waste" (often with the same Hebrew and Greek root) and lands physically described as wilderness or desert are characterized by erratic rainfall patterns, uninhabited land or a land of very limited human activities and presence, and a land that lacks elements for human survival.[4] However, "it can also be used as the locus for healing and moral transformation, where it becomes the locus for spiritual purification"[5] and nourishment for ministry. Although ἔρημος is an ancient word, since the late 1990s it has become a buzz word for the study of the environment, conservation of natural resources, and proper upkeep of forest reserves and the land.[6] In this article, ἔρημος comprises forest reserves, mountains, hills, and any place where human activity is not prevalent and yet it is used for religious purposes periodically.

Significantly, a term whose description comes close to ἔρημος is ὄρος (lit. "*horos*"). It is specifically translated "mountain." It occurs twelve times in the Gospel of Luke (3:5; 4:29; 6:12; 8:32; 9:28, 37; 19:29, 37; 21:21, 37; 23:30). Ὄρος is mainly used to refer to popular and well-known mountains such as Tabor and Olivet or Olives.[7] Mount Olivet was the lodging place for Jesus and his disciples when he entered Jerusalem for the passion. It suggests that it is not too far from urban dwellings. According to Colin Brown, it overlooks the Temple.[8] Both "ἔρημος" and ὄρος express an ancient Semitic idea of mountains, wilderness, and desert, serving as the home of

---

1. Verbrusse, *New International Dictionary*, 206.
2. Funk, "Wilderness," 205.
3. Verbrusse, *New International Dictionary*, 206.
4. Stanley, "Beyond the Campfire's Light," 9–24.
5. Najman, "Toward a Study," 100.
6. Haila, "'Wilderness' and the Multiple Layers," 130.
7. Brown, "ὄρος," 1009–13.
8. Brown, "ὄρος," 1010.

the Supreme Being and gods, particularly Baal.⁹ That notwithstanding, the study confines itself to "ἔρημος" due to its wider references, providing better potential to allow me to draw inferences for the care of the land in Ghana. Most importantly, the care of mountains, hills, natural resources, and forests for religious purposes will directly promote good environmental practice. It is not only for the environment and natural resources, but correspondingly the increasing growth of charismatic Christianity in Ghana, whose leaders give premium attention to praying in the "ἔρημος" such as Aburi Mountain, Legon Gardens, Achimota Forest, Atwea Mountain, etc., that makes a study of this nature quite critical and stimulating.

There are various views concerning ἔρημος in the Gospel of Luke. Two schools of thought are worth mentioning: (i) ἔρημος as a place for eschatological covenant renewal between Israel and YWHW; and (ii) ἔρημος as a location for temptation by Satan. Recently, Clint Burnett observes that Luke engaged ἔρημος as an Old Testament prophetic concept to introduce John as a prophet in the class of Moses and Elijah. It is to fulfill the eschatological restoration of Israel in the wilderness (Ezek 20:35–44). Hence, ἔρημος is a venue for the commissioning of prophets by YHWH and the fulfillment of eschatological restoration/renewal of the covenant relationship between Israel and YHWH.¹⁰ G. H. P. Thompson discussed the temptation account of Jesus and argued that ἔρημος, in the context of the temptation of Jesus, is an intertext of Deuteronomy 8:5 and 9:9. He compared the forty years of Israel's sojourn in the wilderness and the forty days fasting and temptation of Jesus to posit that "ἔρημος" is a place for the temptation of God's servants.¹¹ It is plausible that ἔρημος could be a place for nurturing prophets as postulated by Burnett. However, it is also a place for theophanic revelations for anyone who desires to encounter God through prayer or ascetic life.

In this paper, I use rhetorical exegesis to explore various uses of ἔρημος in the Gospel of Luke. My procedure is to fleetingly discuss the equivalent of ἔρημος in the Old Testament and during the intertestamental period, examine and categorize the various uses of ἔρημος in the Gospel of Luke, discuss its representation by African traditional religionists, assess Pentecostal/charismatic Christian leaders' views of such a representation, outline the implications for Lukan studies on prayer venues, propose the care of such places in Ghana, and then draw some conclusions.

---

9. Verbrusse, *New International Dictionary*, 418.
10. Burnett, "Eschatological Prophet of Restoration," 10–11.
11. Thompson, "Called-Proved-Obedient," 2.

Ἔρημος" in the Old Testament and During the Intertestamental Period

In the Old Testament (and also Septuagint [LXX]), it is a general theological idea to locate the venue for Israel's encounter with YHWH in sacred places.[12] However, YHWH cannot be limited to a particular geographical location as can the gods of the heathens. In Exodus 18–23, Mount Sinai/Horeb was not an ordinary mountain, but a place considered to be a house of God due to the numerous theophanies that Israel encountered. Israel's forty-year-sojourn from Egypt to Canaan was characterized by the revelation of YHWH in the wilderness (LXX: "ἔρημος").[13] It virtually became a norm that once a year, during the Feast of Tabernacles, Israelites would live in tents as if they were in the wilderness (Lev 23:34–36; Deut 16:13–17). Elijah was fed in the wilderness by an angel and was reempowered for ministry on the mountain of Horeb (1 Kgs 19:4–6). The Rechabites deserted their initial place of abode to live in tents in the wilderness (Jer 35:7–9) in order to live close to YHWH. All these signify the importance of the wilderness in the religiosocial life of Israel.

The wilderness was also a place for the abode of demons and vulnerability.[14] It was a place of deadly danger, of separation from God, and of demonic powers (Deut 8:15; cf. Num 21:4–9; Isa 30:6). The scapegoat is chased into the wilderness (Lev 16:10, 21–22). The tension between cultivated land and the wilderness becomes the contrast between the land of Canaan and Egypt; "Egypt can be used specifically as a synonym for the (demonic) desert (Tob. 8:3; cf. Ezek 20:36)."[15] Cities could be said to be chaotic and evil due to economic pressures and their centripetal and centrifugal nature of drawing all manner of people and activities. The inherent chaos found in such cities influenced some to lead ascetic lives so they could draw closer to God. However, it does not suggest that city dwellers are not religious.

During the intertestamental period, the wilderness or desert was a venue for ascetic life. The Essenes believed that the Jewish society was corrupt. Subsequently, they left human society to settle in the desert of Qumran (near the Dead Sea) to establish a monastic community believed to be a pious and holy community for YHWH. It is believed that John the Baptist was a member of this community.[16] The move to settle in the desert was due to

---

12. Funk, "Wilderness," 205.
13. Böcher, "ἔρημος," 1005.
14. Schmutzer, "Jesus Temptation," 21.
15. Böcher, "ἔρημος," 1005.
16. Ayegboyin, *Synoptics*, 34.

Jewish eschatological belief that the Messiah will be revealed in the desert. Hence, they were to prepare the way for him.[17]

"Ἔρήμος" in the Gospel of Luke

In this section, I discuss "ἔρήμος" in Luke 1:80; 3:2, 4; 4:1; 5:16; 6:12; 7:24; 8:29; 9:12; and 15:4. The term ἔρήμος first occurs in Luke 1:80. In comparison with the other canonical gospels, Luke gave appreciable detail concerning the boyhood of John the Baptist. The author of Luke used the anarthrous ἔρήμος to describe the place where John the Baptist was brought up (Luke 1:80). The use of ἔρήμος by the author of the Gospel of Luke follows its use in the LXX. It is a general consensus among Lukan scholars that the author's reference to earlier Scriptures follow that of the LXX. The narratives of the song of Mary, the prophecy of Zachariah, Jesus presented in the Temple, and the statement of Simeon follow a Semitic literary style of the LXX.[18]

In the LXX and Luke, ἔρήμος is usually used with the article in the feminine dative. Luke 1:80, however, is an exemption and that is difficult to explain.[19] It may partially be conjectured that (i) Luke might be using a version of the LXX that is extinct; (ii) he had access to some other Scriptures in addition to the LXX; (iii) it is his own contribution to the LXX version; or (iv) it is information from a proto-gospel or an eyewitness. This is based on Luke's methodology and sources of information that he stated in his Hellenistic literary-styled prologue (Luke 1:1–4).[20]

Prior to the era of Jesus, messianic concepts were at their peak. This gave rise to ascetic life due to the belief that the Messiah will be revealed in the ἔρήμος and will restore the covenant relationship between YHWH and Israel (Ezek 20:35–44). "There thus arises the belief that the last and decisive age of salvation will begin in the ἔρήμος, and that here the Messiah will appear."[21] Hence, those who believe in the Messiah ought to remain in the ἔρήμος until the Christ appears and ends the works of Satan.[22] John the Baptist was born in this religious environment where ἔρήμος had assumed a critical role as the abode of the Messiah, comparable to the role of the Temple in Jewish religious and social life.

17. Böcher, "ἔρήμος," 1006.
18. Litwak, "Use of the Old Testament," 150.
19. Funk, "Wilderness," 214.
20. Just, *Concordia Commentary*, 35–37. See also Crowe, "Sources for Luke and Acts," 75–76.
21. Tübingen, "ἔρήμος," 659.
22. Tübingen, "ἔρήμος," 659.

In Luke 1:80, the author presented ἔρημος as the social context where John the Baptist was brought up. Luke did not indicate that he lived with his priestly parents. The *protevangelium* of James 4:3, an apocryphal document, shows that Herod had wanted to kill John the Baptist, and so Elizabeth, the mother, took him to the ἔρημος of Judea. Elizabeth later died and angels took care of him (John) in the ἔρημος until the time for his public ministry.[23] Leon Morris[24] hypothetically argues that due to the advanced age of the parents of John, they were either unable to care for him or they died and could not nurture John. Hence, he was brought up by the ἔρημος sect of the Essenes because the Essenes voluntarily adopt and train other people's children.

The plural ἐν ταῖς ἐρήμοις (lit. "*en tais herēmos*"), however, does not strongly suggest that John was brought up by the Essene sect near Qumran, although he might have heard about them.[25] It is probable that John was wandering in the "ἔρημος" area traditionally known as Ain el-Maʿmudiyyeh, near Hebron.[26] Others argue for En Kerem, a rocky area in the hills of Judea, where archeologists found a cave believed to be the abode of John the Baptist because a dress made of camel's hair was found in the cave.[27] However, it is too simplistic and an over-generalization to argue that the attire found belongs to John the Baptist without a paleographical dating and analysis test because during the period, there were many messianic adherents in the "ἔρημος," and people living in the ἔρημος were not expected to wear normal attire. It can be summed up that in 1:80, "there is possibly the thought of the wilderness as the place of God's presence."[28] Since ἔρημος is traditionally considered as a spiritual home of prophets in the Old Testament (LXX), Luke's use of ἔρημος is to show that John the Baptist was a prophet from birth, which agrees with the prophecy of the angel Gabriel to Zachariah in the Temple that the child would be filled with the Holy Spirit from birth (Luke 1:11–25). The story is fashioned according to 1 Samuel 3:19–20. In 1:80, he became strong in the spirit (human strength aided by the Holy Spirit).[29]

The author of Luke leaves the narrative of John the Baptist in 1:80 for the story of the birth of Jesus and then goes back to continue John's narrative. Luke has a rhetorical tradition of giving a hint of a person or an issue

23. Moxom, "Boyhood of John the Baptist," 459–61.
24. Morris, *Tyndale New Testament Commentaries*, 89–90.
25. Nolland, *Luke 1-9:20*, 90.
26. Marshall, *Gospel of Luke*, 76.
27. Murphy-O'Connor, "Sites Associated with John the Baptist," 259.
28. Marshall, *Gospel of Luke*, 95.
29. Nolland, *Luke 1-9:20*, 90.

early and later giving full details. Examples include the measly reference to Barnabas in Acts 4:36–37 and later giving more detail about him in Acts 13. Paul was teasingly mentioned in Acts 7:58—8:3, and was given more attention from chapter 9 to 28.[30] In Luke 3:2, a popular phrase that is usually used to describe a spontaneous voice of YHWH to a prophet, is used when John the Baptist began his ministry—"It happened that the word of God came to John, son of Zachariah, in the wilderness." This statement connects with 1:5 and 1:80, and places John among prophets of the Old Testament, particularly Moses and Elijah (Exod 3:1–22; 19; 1 Kgs 19:4–6). The image of "ἔρημος" as a venue for the call of prophets in the Old Testament is well attested.[31] The combination of the "word of God" (ῥῆμα θεοῦ [lit. *rhēma theo*]), "the wilderness" (τῇ ἐρήμῳ [lit. *the herēmō*]), and the mention of key political and religious figures—Pontius Pilate and Herod, Annas and Caiaphas—suggest that the message that John received in the "ἔρημος" was to address issues in their administrations.[32]

The voice of John is reported to have been heard in the ἔρημος (Luke 3:4). It is a reference to Isaiah 40:3 that the Qumran community interpreted as preparing the ἔρημος for the coming of the Messiah. In the Isaiah text, the person behind the voice was not identified. Luke extensively appealed to a well-known and influential text of the period to support the ministry activities of John. An expanded version of Mark 1:2–3 made it specific and categorical that John was sent into the ἔρημος to prepare the way for the coming of the Messiah. In this text, ἔρημος has been presented by the author of Luke as a venue where the voice of God is expressly present and prophets are commissioned.

In Luke 7:24, Jesus described John the Baptist in a manner that closely suggests that he spent his life from childhood to the present stage in the "ἔρημος." His dress code indicates the dress of someone who did not live in a society highly dominated by human activities. It is here confirmed that John is a prophet and more than a prophet. It indicates an inseparable relationship between a prophet and the ἔρημος.

After the baptism of Jesus, the Spirit of God took him into the ἔρημος to be tempted by the devil. Among the numerous people that John baptized, it is only Jesus who experienced theophany and was led to the ἔρημος afterwards. It may be due to his unique ministry assignment. One would expect that the Holy Spirit would lead Jesus into the Temple or synagogue, places specifically designated for worship activities. However, it is obvious that

---

30. Karris. "Gospel According to Luke," 691.
31. Craddock, *Luke*, Interpretation, 47.
32. Just, *Luke 1:1–9:50*, 148.

Jesus was not sent into the ἔρημος to worship but to be tempted by the devil. The phrase ἐν τῇ ἐρήμῳ is an allusion to Deuteronomy 8:2, which concerns how Israel was led into the ἔρημος and tempted.[33] Ἔρημος here is depicted as the habitat of evil spirits. This applies to Luke 8:29 as well.

While the fame of Jesus spread wide and far, he often withdrew into the ἔρημος to pray. Prayer is a theme in the Gospel of Luke and Jesus found the ἔρημος as a venue for prayer (Luke 6:12). It "provides evidence for the idea of mountainous places being near to God; the motif may be drawn from Mark 6:46."[34] Jews like to pray on high-level locations including rooftops and secluded places.[35] Ἔρημος or "Mountain is here a place of retreat to be with God."[36]

Jesus' parable concerning the lost sheep in 15:4 suggests that "ἔρημος" is a place where one can hide valuable items or even hide him/herself to undertake an assignment. The shepherd left the ninety-nine sheep in the ἔρημος in order to go out to look for the lost one. It may be that the shepherd was overtaken by the loss of one sheep and hastily or unconsciously left the ninety-nine in the ἔρημος, as argued by Marshall.[37] However, it remains that ἔρημος housed the ninety-nine sheep. In sum, Luke presents ἔρημος as a place where (i) the presence of God is active and sought for; (ii) prophets are raised; (iii) demonic activities abounds; and (iv) one can hide to undertake a task. The term was used in relation to the ministry activities of John the Baptist and Jesus. It was theologico-rhetorically used to draw John the Baptist and Jesus into the prophet catenae of the Old Testament. Theologically, Luke adapts the use of the term in the LXX to demonstrate that the presence of God, as experienced by key prophets in the Old Testament such as Moses and Elijah, was still present at ἔρημος and it endows John and Jesus for ministry. Luke's use of the Old Testament is more akin to the LXX than the Masoretic text.[38]

Rhetorically, Luke used ἔρημος four times in relation to John the Baptist (1:80; 3:2, 4; 7:24). It was used three times in relation to Jesus (4:1; 5:16; 6:12). Repetition is a rhetorical device employed to attract the attention of listeners or readers to the argument of the author. Depending on the circumstances or the cause that merit the situation, repetition can be progressive (positive), and regressive (negative). A conclusion may be based on at least

33. Marshall, *Gospel of Luke*, 169.
34. Marshall, *Gospel of Luke*, 237.
35. Aryeh, "Exegetical Discussion of Mark 2:1–12," 9.
36. Nolland, *Luke 1–9:20*, 269.
37. Marshall, *Gospel of Luke*, 601.
38. Adams, "Genre of Luke and Acts," 97–120.

two prior premises: major and minor in a deductive argumentation. Repetition always make reference to the past in new forms and contexts. In other words, it is the restatement of what has been stated earlier. Neil R. Leroux refers to both progressive and regressive repetition as syllogistic progression form.[39] In this situation, the repetitive use of ἔρημος is progressive. It was initially used to describe the place for the nurture, training, and abode of John the Baptist (1:80). Secondly, it was used to describe the ministry jurisdiction of John the Baptist (3:2, 4), which is an allusion to Isaiah 40:3 and Malachi 3:1. The intention was to appeal to an established Scripture to buttress the ministry activities of John as a prophet, and to the fulfilment of prophecy.[40] Thirdly, it was used by Jesus to confirm John as prophet (7:24). Obviously, any statement of Jesus carries heavy weight. However, it is not clear what the phrase "more than a prophet" means.

Luke presents ἔρημος as a place for raising prophets. It emphasizes the notion that John was the last prophet of the Old Testament. Walter C. Kaiser Jr. argues that John the Baptist is the fulfillment of the second coming of Elijah and that he was not the only forerunner to the Messiah, but that he was one of the forerunners in the series of forerunners in history.[41]

The use of ἔρημος in relation to Jesus in all its four occurrences was to consistently demonstrate that it is a place to pray and receive power for ministry (4:1, 14; 5:16), and a place to hide valuables (15:4). Luke demonstrates that effective prayer leading to theophanic manifestation does not often take place in the Temple or synagogue. By this postulation, he and his audience, being gentiles, suggest that the presence of God is not limited to the Temple. It can be experienced even at "ἔρημος," a place where Israelite prophets experienced YHWH prior to the building of the Temple and afterward. However, rabbinic tradition suggests that an angel could appear to the high priest in the Holy of Holies during the feast of Yom Kippur.[42] Hence, Luke used "ἔρημος" to emphasize the Old Testament idea of a place to encounter God.

The Use of Wilderness Motif in Ghana

The use of wilderness in Ghana is varied. It is used by both the Christian community and practitioners of African Traditional Religion (ATR). In the context of ATR, wilderness refers to sacred places: hills, mountains, valleys,

39. Leroux, "Repetition, Progression, and Persuasion," 1–25.
40. Beale, *Handbook on the New Testament*, 56.
41. Kaiser, "Promise of the Arrival," 221–33.
42. Elgvin, "From the Earthly," 25.

forests, large rocks, and groves.[43] In the Christian context, it is used to refer to grottoes, and prayer camps outside church auditoria.

The Concept of Sacred Space in African Traditional Religion

In ATR, the sacred space concept is a set apart spatial area of varied size earmarked for religious activities. John S. Mbiti states that sacred places "are symbolically the meeting-point between the heaven or sky and the earth, and therefore of the visible and invisible world . . ., people regard them as sacred and therefore as places where they feel the symbolic presence of God."[44] It is a place that the spirits have chosen as their habitat. Shrines at these locations houses religious symbols, objects, and artifacts of various shapes and sizes that also serve as the religious ideas and concepts of a community. It is a place that helps to mediate the presence of the spirits through offerings, sacrifices, and libations, among other things. Common human activities such as hunting, cutting wood for fire or building, and farming were not allowed in or around sacred places.[45] Animals that run into such places were not supposed to be hunted; because they may be spirits that have manifested in animal forms. Mbiti intimates that "Hills and mountains lift people's eyes toward the sky, and hence toward God and the heavenly world. For that reason many of them all over Africa, including the continent's highest mountains, Mt Kilimanjaro, Mt Kenya, Mt Elgon, the Ruwenzori (or mountain of the moon), Mt Cameroon, and others, are natural religious monuments for the people who lived nearby."[46]

Although there are shrines in some homes in African communities, the shrine at the mountains were patronized on a mass basis. It serves as church building, cathedrals, and mosques for ATRs because it is mostly owned by the entire community. The encounter between humans and the spirits in ATR is special and mostly mediated at such spaces. It suggests that some human activities can be a barrier to divine encounters between the deities and its adherents. There is the belief in some African communities that deities and spirits that were not well cared for leave to reside in the mountains and are rediscovered in those venues that have been declared as sacred places. Since ATR does not discriminate by asking people to denounce their erstwhile religious affiliations before accessing the services of

43. Okyere, "Reconstructing Sacred Places," 3.
44. Mbiti, *Introduction to African Religion*, 149.
45. Mbiti, *Introduction to African Religion*, 150.
46. Mbiti, *Introduction to African Religion*, 152.

traditional priests or priestesses, a mountain serves as a cover for pilgrims who adhere to other religions and would not like to be seen by others at shrines. It is important to state that, environmentally, human activities may have some negative effect on the wilderness. However, its taboos of not permitting hunting and not cutting down trees around such sites help to protect endangered species and trees.[47]

It is significant to mention that ATR is gradually losing its "wilderness" concept, probably due to modernism and globalization. One of the largest and most popular traditional sacred places in Ghana is located at Forikrom, a rural community situated in the Techiman Municipal Area in the Brong Ahafo Region. This sacred place houses the Bote shrine, magic cave and holy mountains. The most important religious event in the community is the Bakor rituals (traditional bath rituals). It attracts tourists and ATRs who come to sacrifice and give offering to the spirits for favor. These sacred places have been left to deteriorate and finally be abandoned by the community due to nonobservance of traditional rules regarding sacred places by the youth, who do not regard the sacredness of those places. This could be due to the lack of sacred Scriptures in traditional religion which can serve as a guide for the youth.[48] Similarly, Tanchara, a farming community located between two hills/rocks in the Lawra District of the Upper West Region seems to be disregarded. The community had a population of 3,800 people as of 2010. It had seventy sacred places in 2007. This was reduced to twenty-five in 2009.[49] This rapid decrease in the number of sacred places could be due to the need for hardwood for building and other purposes, unregulated mining, and quarry sites for economic reasons.[50]

Wilderness as Sacred Space in Ghanaian Christianity

In the Christian religious milieu, wilderness serves as a venue for ascetic life, prayer and fasting, healing, and the revival of religiosocial life for many Ghanaians. The Roman Catholic Church has grottoes and wilderness places for prayer, meditation, and study. According to J. Kwabena Asamoah-Gyadu, St. Mary's Sanctuary (SMS), located at Buoho, Kumasi, in the Ashanti Region, is one of the officially recognized Roman Catholic grottoes in Ghana.

---

47. Shen et al., "Tibetan Sacred Sites," 9.

48. Aryeh, "Role of Ascetism," 153.

49. 2010 Annual Report of Centre for Indigenous Knowledge and Organizational Development.

50. Sheridan, "Environmental and Social History," 75.

It functions under the auspices of the Konongo Mampong Diocese.[51] The SMS was established in the early 1940s by Rev. Fr. Kwaku Tawiah Yesereh, a Roman Catholic priest. The term "Buoho" literally means "besides the rock or stone/hill." The venue was initially a traditional sacred place for spirits and deities of ATR, which was later taken over by the Roman Catholic Church and became an important Christian center for pilgrimage, prayer, and retreat. This means that either traditional religion is gradually losing its sacred places to Christianity, or the exclusivity of "ἔρήμος" in traditional religion is losing its significance.

In the act of taking over traditional sacred places, Christianity have found expression in traditional sacred places.[52] One of the Christian sacred spaces that has received a lot of attention are the Abasua Mountains, popularly referred to as the Atwea Mountains, situated in the Ashanti Region. History has it that the site was a traditional sacred space for an *Akan* deity; the venue is one of the four mountains in Abasua called *Atwea bɔɔ*, known for protection and giving answers to landmark questions.[53] There are two claims to the Christian discovery of Abasua Mountain. Okyere argues that the site was discovered for Christian use by the Seventh-Day Adventists (SDA) in 1958. He explained that some members of the church, who resettled in Abasua from Asante Agona and Adutwam, started the church in Abasua. The pioneer members of the church were adherents and priests or priestesses of the *Atwea bɔɔ* shrine, and therefore led some Christians to the mountain to pray.[54] Asamoah-Gyadu also postulates that the Abasua Mountain was discovered for Christian use by the late Reverend Abraham Osei-Asibey, a Methodist minister, as a place for prayer in 1963, after he had revelations directing him to go there to pray.[55] On the first visit to the mountain by Reverend Osei-Asibey and his *Kristo Mu Anigye Kuo* (lit "Happiness in Christ Fellowship"), a prayer and evangelism team, to pray, they claimed to have experienced the descending of the clouds on the mountain, which is similar to that of the biblical Mount Sinai. It is likely that after the initial discovery of the mountain in 1958 by the SDA, it was abandoned and later rediscovered by Reverend Osei-Asibey in 1968.

During my visit to the mountain in 1999, I met a lot of newly ordained young pastors mainly from Pentecostal and charismatic ministries praying

---

51. Asamoah-Gyadu, "On the 'Mountain' of the Lord," 65–86.
52. Obeng, *Asante Catholicism*, 161.
53. Okyere, "Reconstructing Sacred Places," 40–44.
54. Okyere, "Reconstructing Sacred Places," 47.
55. Asamoah-Gyadu, "On the 'Mountain' of the Lord," 70–73.

for the requisite anointing[56] to go and begin ministry. The only building on Abasua Mountain was a Methodist chapel, and some temporary tents. During my last visit in 2016, I realized that many permanent structures have been built on the Abasua "ἔρημος." It is an indication that Abasua Mountain is increasingly patronized by large number of Christians from all over Ghana, hence the need to care for sacred spaces in Ghana.

Implications

It has become obvious from the above discussions that wilderness plays a significant role in the religious lives of ancient Israel, Luke and his audience, traditional religion in Ghana, and many Christian denominations in Ghana. It provides the atmosphere for prayer leading to theophanic revelations, healing, and revival. This phenomenon engineers implications for Lukan studies, and the patronage of sacred spaces in Ghana.

First, M. A. Powell asserts that worship and prayer is one of the major themes discussed in the Gospel of Luke, thus, the book opens and closes with people worshiping and praying in the Temple (1:8; 24:53).[57] He seems to limit prayer to the Temple. This postulation would have to be seen in the light of the use of wilderness for prayer and theophanic experiences by Jesus and John the Baptist. The emphasis on the Temple to the non-Jewish audience of Luke has the potential of creating a proselytic agenda of converting the heathen to Judaism rather than consolidating the faith of Theophilus in Christianity, as stated in the prologue (1:1–4).

In the context of Theophilus, wilderness is conceived as a lonely place where the deities reveal themselves to their subjects through an intermediary priest.[58] It is contextually apt to theologize the substitution of the deities to God (Almighty) on mountains, thereby making the presence of God available and felt at all locations, including the mountains of non-Jewish people. Clearly, the gospel opens and closes with people praying in the Temple as argued by Powell. However, John the Baptist and Jesus also prayed at the mountains. The Temple is dedicated to priestly activities, but the wilderness has been democratized for use by anyone who seeks to encounter God personally. Luke adopted the use of wilderness in the Old Testament (LXX) and during the intertestamental period where messianic interest was at its apex.[59]

56. "Anointing" here refers to the manifestation of the gifts of the Spirit recorded in 1 Cor 12:1–8.
57. Powell, *Introducing the New Testament*, 157–58.
58. Verbrusse, *New International Dictionary*, 206.
59. Funk, "Wilderness," 205–7.

Second, Africans are particularly religious and issues of religion are critical. In Africa, "religion is life and life religion."[60] Religion permeates every fiber of societal norms. This is quite similar to the concept of "ἔρημος" in the LXX/Old Testament and Luke, with ATR referring to it as a sacred place dedicated to the deities. It is a taboo to farm, hunt, cut trees, mine, and set up a quarry around sacred places. Rocks wre not to be mined because they were the residences of deities. Trees were also considered as residences of the deities, hence, they were not to be cut down.

In view of environmental degradation in Ghana, and the recent outcry to protect natural resources and regulate *galamsey* (illegal small-scale mining), I posit that a body (organization) be set up to identify mountains in Ghana for religious (Christian and ATR) activities. These mountains should be managed by a particular religious authority. Allocation can be done through tendering processes that should include proposals to maintain the "ἔρημος." It is hoped that when the religious functions of mountains are emphasized by government, traditional leaders, and the church, natural resources will not be at the mercy of human activities in the name of the quest for economic well-being.

Conclusion

In this paper, I attempt to argue that ἔρημος as wilderness, mountain, and sacred space provides the venue to encounter the presence of God, a place where demonic activities abound, and a place where traditional shrines are set up for a community. I analyzed the theologico-rhetorical use of ἔρημος in the Gospel of Luke concerning John the Baptist and Jesus, and discussed its role in the religious and social lives of ATRs. I likewise examined the use of mountains and sacred spaces by various Christian denominations in Ghana to support my postulation and have drawn implications for Lukan studies on prayer, and for the care of the land in Ghana. In view of the significant role that sacred spaces played in the training of John the Baptist and the prayer life of Jesus, one can liken it to a place for prayer and can be compared to the Temple. This resonated quite well with the non-Jewish audience of Luke, who were not familiar with the Temple and its regulations.

The religious nature of African society makes mountains become a central location for communal religious activities. As a "sacred place" this shows the degree to which it is held in high esteem. The taboos associated with these spaces clearly will help the protection and conservation of natural recourses. Hence, the call for the protection of the environment and the conservation of

---

60. Opoku, *West African Traditional Religion*, 1.

natural resources should not be limited to the tourist and biodiversity benefits of sacred spaces, but must include its religious benefits.

Bibliography

Adams, Sean A. "The Genre of Luke and Acts: The State of the Question." In *Issues in Luke-Acts: Selected Essays*, edited by Sean A. Adams and Michael W. Pahl, 97–120. New York: Gorgias, 2012.

Aryeh, Daniel Nii Aboagye. "The Role of Asceticism and Sacred Places in Ministerial Formation in Ghanaian Christianity: Exegetical Analysis of Luke's Theology of the Temptation of Jesus (4:1-13)." *Journal of Applied Thought* 5.1 (January 2016) 137–62.

———. "An Exegetical Discussion of Mark 2:1–12: Lessons for Forgiveness and Healing in Contemporary Christianity in Ghana." *Conspectus* 25 (March 2018) 1–20.

Asamoah-Gyadu, J. Kwabena. "On the 'Mountain' of the Lord: Healing: Pilgrimages in Ghanaian Christianity." *Exchange* (2007) 65–86.

Ayegboyin, Deji. *The Synoptics: Introductory Notes on the Gospels According to Matthew, Mark and Luke*. Ibadan, Nigeria: Greater Height, 2004.

Beale, G. K. *Handbook on the New Testament Use of the Old Testament: Exegesis and Interpretation*. Grand Rapids: Baker Academics, 2012.

Böcher, O. "ἔρήμος." In *The New International Dictionary of New Testament Theology*, edited by Colin Brown, 1005–8. Grand Rapids: Zondervan, 1971.

Brown, Colin. "Ὄρος." In *The New International Dictionary of New Testament Theology*, edited by Colin Brown, 1009–13. Grand Rapids: Zondervan, 1971.

Burnett, Clint. "Eschatological Prophet of Restoration: Luke's Theological Portrait of John the Baptist in Luke 3:1–6." *Neotestamentica* 47.1 (2013) 1–24.

Craddock, Fred B. *Luke, Interpretation: A Bible Commentary for Teaching and Preaching*. Louisville: Westminster John Knox, 1990.

Crowe, Brandon D. "The Sources for Luke and Acts: Where Did Luke Get His Materials (and Why Does It Matter)?" In *Issues in Luke-Acts*, edited by Sean A. Adams and Michael W. Pahl, 73–85. New York: Gorgias, 2012.

Elgvin, Torleif. "From the Earthly to the Heavenly Temple: Lines from the Bible and Qumran to Hebrews and Revelation." In *The World of Jesus and the Early Church: Identity and Interpretation in Early Communities of Faith*, edited by Craig A. Evans 23–36, Peabody, MA: Hendrickson, 2011.

Funk, Robert W. "The Wilderness." *Journal of Biblical Literature* 78.3 (Sept. 1959), 205–14.

Haila, Yrjö. "'Wilderness' and the Multiple Layers of Environmental Thought." *Environment and History* 3.2 (June, 1997) 129–47.

Just, Arthur A., Jr. *Concordia Commentary, A Theological Exposition of Sacred Scripture: Luke 1:1–9:50*. St. Louis: Concordia, 1996.

Litwak, Kenneth D. "The Use of the Old Testament in Luke-Acts: Luke's Scriptural Story of the 'Things' Accomplished among us." In *Issues in Luke-Acts*, edited by Sean A. Adams and Michael W. Pahl, 147–69. New York: Gorgias, 2012.

Kaiser, Walter C., Jr., "The Promise of the Arrival of Elijah in Malachi and the Gospels." *Grace Theological Journal* 3.2 (1982) 221–33.

Karris. Robert J. "The Gospel According to Luke." In *The Jerome Biblical Commentary*, edited by Raymond E. Brown et al., 673–721. London: Burns and Oates, 2007.

Leroux, Neil R. "Repetition, Progression, and Persuasion in Scripture." *Neotestamentica* 29.1 (1995) 1–25.

Marshall, I. Howard. *The Gospel of Luke: A Commentary on the Greek Text*. Grand Rapids: Eerdmans, 1987.

Mbiti, John S. *Introduction to African Religion*. 2nd ed. Oxford: Heineman, 1991.

Morris, Leon. *Tyndale New Testament Commentaries: Luke*. Rev. ed. Leicester: Inter-Varsity, 1988.

Moxom, Philip S. "The Boyhood of John the Baptist." *The Biblical World* 10.6 (Dec. 1897), 454–61.

Murphy-O'Connor, Jerome. "Sites Associated With John the Baptist." *Evue Biblique* 112.2 (April 2005), 253–66.

Najman, Hindy. "Toward a Study of the Uses of the Concept of Wilderness in Ancient Judaism." *Dead Sea Discoveries* 13.1 (2006) 99–113.

Nolland, John. *Word Biblical Commentary: Luke 1-9:20*. Dallas: Word, 1989.

Obeng, Pashington. *Asante Catholicism: Religious and Cultural Reproduction among the Akan of Ghana*. Leiden: E. J. Brill, 1996.

Okyere, Philip Kwadwo. "Reconstructing Sacred Places: The Place and Relevance of Abasua Prayer Mountain in Contemporary Ghanaian Christianity." MPhil Thesis, University of Ghana, Legon, Ghana, 2012.

Opoku, Kofi Asare. *West African Traditional Religion*. Accra, Ghana: FEP International, 1978.

Powell, Mark Allan. *Introducing the New Testament: A Historical, Literary, and Theological Survey*. Grand Rapids: Baker Academic, 2009.

Schmutzer, Andrew. "Jesus Temptation: A Reflection on Matthew's Use of Old Testament Theology and Imagery." *Ashland Theological Journal* (2008) 15–42.

Shen, Xiaoli, et al. "Tibetan Sacred Sites Understanding the Traditional Management System and Its Role in Modern Conservation." *Ecology and Society* 17.2 (June 2012) 1–12.

Sheridan, Michael J. "The Environmental and Social History of African Sacred Groves: A Tanzanian Case Study." *African Studies Review* 52.1 (Apr. 2009) 73–98.

Stanley, George H. "Beyond the Campfire's Light: Historical Root of the Wilderness Concept."

Natural Sources Journal 29.1 (Winter 1989) 9–24.

Thompson, George H. P. "Called-Proved-Obedient: A Study in the Baptism and Temptation Narratives of Matthew and Luke." *The Journal of Theological Studies* 11.1 (April 1960) 1–12.

Tübingen, Gerhard Kittel. "Ἔρημος." In *Theological Dictionary of the New Testament, Vol. II*, edited by Gerhard Kittel, translated by Geoffrey W. Bromiley, 659–61. Reprint. Grand Rapids: Eerdmans, 1982.

Verbrusse, Verlyn D., ed. *New International Dictionary of New Testament Theology*. Abridged ed. Grand Rapids: Zondervan, 2000.

Salvaging our Environment: A Reflection on Responses of the Catholic Church in Ghana

BONSU OSEI-OWUSU

Methodist University College, Accra

Abstract

One of the greatest contemporary matters requiring an informed missional response is the environmental crisis. Because of a growing consensus that Ghana may be moving toward an unprecedented ecological disaster, the Christian community in the country should reexamine some of its theological assumptions and fill in some gaping blind spots in its understanding of the Missio Dei. This paper examines the extent to which the institutional environmental policies, programs, practices, and theologies of the Catholic Church in Ghana are being translated into the life patterns of its members. The paper blends environmental ethics and religion as it adopts qualitative techniques in data collection and analysis. It uses a series of group interviews to gather data from eight selected congregations of the Catholic Church in Ghana to assess the attitudes toward, and actions regarding, the environment amongst members. It suggests that despite a wealth of institutional resources that have been developed by the church to foster theologically informed environmental knowledge, very little awareness of the institutional church's ethics is found in the local church context. The church therefore must reorient itself to adequately improve upon its efforts at environmental sustainability because the church's mission involves both humanithy's and nature's redemption.

Introduction

IN GENERAL, ENVIRONMENTAL ETHICS presents and defends a systematic and comprehensive account of the moral relations between human beings and their natural environment. Environmental ethics assumes that human behavior toward the natural world can be and is governed by moral norms. A theory of environmental ethics then must go on to (1) explain what these norms are; (2) explain to whom or to what humans have responsibilities; and (3) show how these responsibilities are justified. A variety of environmental ethics emerge with these explanations. Perhaps one explanation is that our responsibilities to the natural environment are only indirect—that the responsibility to preserve resources, for example, is best understood in terms of the responsibilities that we owe to other humans. Another may be that we have direct responsibilities not only to plants and animals but also to ecosystems and species, and that these responsibilities are based on the moral standing of these natural objects.

An understanding of Christian mission must respond to issues and questions that arise in living our faith. Theology is not formed in a vacuum, but emerges in response to concrete situations or crises that stimulate study and reflection. The issues shaping our approach to mission today are different from those our ancestors struggled with, and in all likelihood, will not be cutting-edge concerns fifty years from now. As each new crisis is addressed, however, our perspective on the nature of the Christian mission is enriched and enlarged.

One of the greatest contemporary matters requiring an informed missional response is environmental degradation. Ghanaian communities today are languishing under the weight of enormous environmental problems to the point that they seem not to have any idea how to confront the situation. Environmental problems ranging from choked and dying water bodies, sea erosion, soil infertility, loss of plant and animal species, and a generally filthy environment as a result of indiscriminate waste disposal, are problems most Ghanaian communities face today. Again, because of a growing consensus that the planet may be moving toward an unprecedented ecological disaster, the Christian community should reexamine some of its theological assumptions and fill the gaping blind spots in its understanding of the *Missio Dei*. The church, therefore, in every camp and tradition should be grappling with its responsibility toward creation and seeking to integrate this in its missionary praxis.

This is in line with Paul Collins's observation that our self-definition as human beings is derived from two primal sets of relationships. The first, he believes, is our relationship to one another and to the natural world and all that is in it. The second is our mysterious, intangible, and difficult relation to the transcendent: it is within this context that we sort out our meaning of existence. This he calls spirituality or religion. Collins further observes that the environment is central to the future of religion. He believes religion will gradually cease to exist if the natural world continues to be devastated at the present rate. According to him, there is a dependent interrelationship between the development of religious attitudes and the sustainability of the natural world.[1] The argument of Collins shows that the future of religion (especially Christianity) hangs on the sustainability of the environment and this throws a big challenge to Christians to come out with practical and workable solutions to help salvage our environment.

It is clear from the above that there are very important reasons why the church needs to be concerned about the environmental issues in the country. In a very deep religious sense in Christian understanding, the whole creation is believed to have come from God (Gen 1:1–31). For instance, the sun, the moon, the water, the air, the land, its plants, and all other life forms contribute to making life possible. Humanity is supported by, and dependent on, these living and nonliving elements God created for her continuous existence. The environment has also been entrusted to human beings by the Creator to be its steward. The church therefore has a role to play in the struggle to save the environment from total destruction. According to Pope John Paul II, "the ecological crisis is a moral issue."[2] And since religion and morality are intertwined, Christians are in a vital position of leading the campaign against the destruction of the environment upon which their very survival is dependent.

The paper therefore discusses the role of the church in salvaging our environment. The main focus of this paper is to trace historically and highlight the responses of the Catholic church in Ghana in proposing solutions to the problem the environment. The paper also assesses the various programs and activities of the church that are aimed at helping society fight against the environmental menace in Ghana and the extent to which these policies and programs are being translated into the life patterns of its members.

The Role of the Church in Salvaging Our Environment

The Catholic church in Ghana, in its bid to help address the environmental problem, has come out with programs that include the issuing out of communiqués and the Celebration of Abor Week. (*Abor* is a Latin for tree). Abor

1. Collins, *God's Earth*, 3
2. Squicciarini, *Pope John Paul II's Message*, 2.

Week is therefore a week dedicated to tree planting. The Catholic church in Ghana, like her international counterparts, is promoting environmental care as a religious and moral obligation. In 1980, a portion of the communiqué issued by the Ghana Catholic Bishops' conference centered on environmental crises such as desertification and the wanton destruction of powerless farmers' property. In the communiqué, the bishops lamented the indiscriminate burning of bushes and forests, especially in the northern part of the country, and the wanton destruction of the poor farmers' cocoa and crop farms by timber loggers and road contractors. The bishops appealed to the government through the communiqué to use her powers to put a stop to "bush burning without supervision and the felling of trees without the corresponding afforestation."[3] Excerpts of the communiqué read as follows:

> The conference notes with great concern that there is indiscriminate burning of bush and forest in certain parts of the country, especially in the north which can only hasten the process of desertification. The conference, therefore, appeals to the Government to use its powers to put a stop to bush burning without supervision and the felling of trees without the corresponding afforestation. The Bishops' Conference consider the wanton destruction of poor farmers' cocoa and crop farms by timber and road contractors to be a crime against justice.[4]

Similarly, the communiqué from the 1987 conference complained about the depletion of the country's forests through deliberate cutting of trees and bush burning. Although the conference acknowledged the economic value of timber in Ghana, it appealed to the timber contractors to exercise restraint and circumspection in the felling of trees and to all, especially the youth, to intensify their efforts in afforestation.[5]

Furthermore, the 1995 communiqué of the Catholic Bishops' conference also complained about the destruction done to the environment through surface mining and the technique of heap leaching. They appealed for effective action to minimize, or eliminate, the damage involved.[6] In the 1996 communiqué, the bishops warned against the grave danger in the degradation of our environment through human activities such as the indiscriminate felling of trees for commercial purposes, the constant burning of the bush, and some mining techniques.[7]

3. The Ghana Catholic Bishops' Conference, *Ghana Bishops Speak*, 44.
4. The Ghana Catholic Bishops' Conference, *Ghana Bishops Speak*, 44–45.
5. The Ghana Catholic Bishops' Conference, *Ghana Bishops Speak*, 120.
6. The Ghana Catholic Bishops' Conference, *Ghana Bishops Speak*, 6.
7. The Ghana Catholic Bishops' Conference, *Ghana Bishops speak*, 27–30.

In 1997, the bishops' conference, as a way of demonstrating their commitment to the regeneration and preservation of the environment, formally instituted the first week of June every year as Abor Week—a week which is dedicated to tree planting. To this end, the 1997 communiqué set forth:

> We cannot but revisit our preoccupation with the environment, which has been the subject of our communiqués in the last few years. By way of demonstrating our commitment to the regeneration and preservation of the environment, we hereby institute the first full week of June of every year as *Abor Week* to be observed as the Church's sanitation and environment week. We are aware that there are many laws in place to protect our environment. We plead for their enforcement.[8]

According to N. A. Raphael, in 1925, the then-governor of the Gold Coast (now Ghana), Sir Frederick Gordon Guggisberg (1919–1927), instituted an Abor Day to originally encourage the planting of palm trees. Later however, the planting of neem trees replaced the palm trees to create avenues and parks, and as a sort of a national afforestation venture.[9] Therefore, the Abor Week of the Catholic church in Ghana is recalling this historical environmentally friendly precedent, extending it from one day to one whole week and from only a tree planting concern to cover the total care of our environment.

N. A. Raphael has observed that the rationale behind this Abor Week is to:[10]

i. Re-awaken and conscientize people about their religious and moral obligation and responsibility as stewards of God's creation for the proper care of the earth;

ii. Identify and embark on possible practical solutions or actions to solve the environmental problems in our various communities and homes;

iii. Advocate and seek the enforcement of policies and laws governing the sustainable use and management of our natural resource; and

iv. Examine and judge for ourselves, our lifestyles, sanitary practices and the state of our immediate surroundings and take remedial measures.

Although the Ghana Catholic Bishops' conference reinstituted the Abor Week, according to Rev. Fr. Wisdom Larweh, the Assistant Secretary General of Catholic church in Ghana, the celebration now involves religious bodies or

8. The Ghana Catholic Bishops' Conference, *Ghana Bishops Speak*, 41–42.
9. Raphael, *Abor Week*, 10.
10. Raphael, *Abor Week*, 12.

groups and other organizations such as the Christian Council of Ghana, the Ahmadiyya Muslim Mission, World Vision, and the National Commission on Culture. This is to uphold their religious and moral obligation and responsibility as stewards of God's creation. The celebration is under the auspices of the Environmental Protection Agency.[11]

The National Laity Council[12] of the Catholic church has also taken up the issue of environmental crisis, and for three years, at their Annual National Laity Congresses, have centered their themes on the environment. For instance, according to the Assistant Secretary General, the theme for the 1997 and 1998 congresses were "The Church and the Protection of the Environment" and "The Protection of Our Environment as Our Christian Responsibility," respectively. This was to enable the laity of the church to rise up to their moral and religious responsibility toward the environment, to help create awareness in their communities about the destructive effects of environmental degradation, and to compliment the efforts of their bishops. The Assistant Secretary General opined that, additionally, some of the dioceses have also embarked on agro-forestry and tree planting projects. In the Bongo agro-forestry project, according to the Assistant Secretary General, an average of 120 tree seedlings are planted every year within the area. The church also gave support to fifty farming families in the establishment of family woodlots, and about eight schools in Garu have undertaken tree planting on a competitive basis. These projects are undertaken in the Navrongo-Bolgatanga diocese. The Assistant Secretary General further observed that the Department of Socio-Economic Development (DSED) of the National Catholic Secretariat also undertakes education programs, including seminars for selected farmers in their various areas of operation. These farmers are taught farming techniques and practices which are environmentally friendly. The rationale is to help minimize the harm and destruction their farming activities can cause to the environment.[13]

In 1998, the Bishops' Conference lamented the Akosombo Dam Crisis and appealed to people with the means to embark on afforestation on the banks of the Volta River and lake. According to the bishops, "It is a fact that there is a kind of symbolic relationship between water and trees. Trees depend on water to grow and rain often depends on trees to fall."[14]

11. Interview with Rev. Fr. Wisdom Larweh, The Assistant Secretary General Ghana Catholic Bishops' Conference, on 31st October, 2016 at the National Catholic Secretariat, Accra

12. "Laity" refers to members of the various parishes who are not of the clergy or the priestly order.

13. Interview with Rev. Fr. Wisdom Larweh, March 2017.

14. The Ghana Catholic Bishops' Conference, *Ghana Bishops Speak*, 56–67.

The communique from the 2000 Ghana Catholic Bishops' conference also complained about the destruction done to the environment by timber contractors and surface miners and cautioned all about the dangers ahead if those involved do not "repent" from their activities. A portion of the communiqué reads as follows:

> Our environment is being desecrated at a frightening rate as we have had cause to point out in many communiqués, and that in a few years' time as much as about 35% of the surface area of Ghana could become a desert; and yet we look on powerlessly when our wood is cut for short-term economic gain without it being replaced. Laws governing the re-planting of trees by timber contractors for trees they cut must be enforced rigorously. The same is true about laws to protect our nation. Otherwise our nation is heading for a sure ecological catastrophe.[15]

The 2001 conference likewise touched on the destruction done to the environment by surface miners and bush fires and therefore appealed to the people of the Upper West Region in particular to desist from bush burning and the indiscriminate felling of trees. To this end, the 2001 communiqué had this to say:

> We have in the past bemoaned the catastrophic effect of surface mining and the wanton destruction of our forest. We have been very much touched by the reality that the environmental degradation due to bushfires has taken on a disastrous dimension here in Upper West Region.[16] We must realize that the destruction of the environment leads to poverty, scarcity of rain, depletion of water bodies, and mass migration. We, therefore, appeal to the people of the Upper West Region in particular who appear to be suffering from the worst effects of the change in environment to desist from bush burning, indiscriminate felling of trees especially economic trees, and certain traditional farming methods. If such precautions are not taken, we should not be surprised that the Sahara Desert consumes a greater part of the upper regions of Ghana in the foreseeable future. Accordingly, we renew our repeated appeal for the enforcement of laws governing surface mining and bush burning in Ghana.[17]

The 2004 Ghana Catholic Bishops' Conference Annual Plenary Assembly was held at the Mole National Park Motel in Damongo in the West

---

15. The Ghana Catholic Bishops' Conference, *Ghana Bishops Speak*, 72–73.
16.
17. The Ghana Catholic Bishops' Conference, *Ghana Bishops Speaks*, 80–81.

Gonja District of the Northern Region. From the conference communique, it was observed that the area is endowed with the biggest game reserve in West Africa and also has one of the biggest waterfalls in Ghana. Participants at the conference suggested that these great assets be further developed and preserved in a better condition, and therefore appealed to hunters to desist from hunting game indiscriminately and without a genuine permit. Concern was expressed as to the rapidity with which the people were destroying our forests, lands, and water bodies, and suggested the need to go in massively for afforestation.[18]

In 2007, the Ghana Catholic Bishops' Conference reiterated their constant call for protection of the environment. To this end, the bishops had this to say:

> We once again reiterate our constant call of the protection of our environment. The destruction of the ecology has already had disastrous effects on our lives and yet our forest continues to be depleted. The recent flood disaster in the Northern Regions of Ghana and elsewhere indicate that the irresponsible and unrestrained human activities can have disastrous effects on our environment, human life and property. We believe that environmental catastrophes such as we have experienced recently do come about due to environmental degradation caused by activities such as bush burning, indiscriminate felling of trees, sand wining and surface mining.[19]

The 2010 Ghana Catholic Bishops' Conference Annual Plenary Assembly was held in Sefwi Wiawso under the theme: "If you want to cultivate peace, protect creation." The bishops deliberated on the management of our God-given environment and the exploitation of our mineral and natural resources as well as the impending oil and gas production, which seems to hold promise for all Ghanaians, if well managed. The portion of the 2010 communiqué that dealt with the environment focused on the following thematic areas: respect for creation, negative impact of mining, pollution, deforestation, and undermining our peace and security. The bishops concluded by advising Ghanaians thusly: "Let us always bear it in mind that God has made us stewards of creation for creation to serve our needs. However, if we

---

18. The Ghana Catholic Bishops' Conference, *Ghana Bishops Speaks*, 104–5.

19. Ghana Catholic Bishops' Conference, "Communiqué Issued by the Ghana Catholic Bishops' Conference at its Annual Plenary Assembly held in Christian Village, Kumasi," (2nd—10th November, 2007), 3.

destroy creation, we destroy ourselves. Let us endeavor, in all that we do, to protect creation and thereby safeguard our own future."[20]

The 2013 Ghana Catholic Bishops' Conference Annual Plenary Assembly was held in Sunyani on the theme: "The New Evangelization for the Transmission of the Christian Faith in Ghana." The Catholic bishops, in a communiqué, called on Ghanaians to acknowledge that their role toward the environment, as collaborators with God, is fundamentally one of stewardship and not exploitation. A portion of the 2013 communiqué reads as follows: "Our role as collaborators with God is fundamentally one of stewardship and not exploitation. We have a God-given task to protect and use the environment judiciously and with responsibility for future generations who equally have a stake in the world. The new evangelization requires inculcating this new mindset in all of us as we undertake our socio-economic activities." [21]

The 2016 Ghana Catholic Bishops' Conference Annual Plenary Assembly was held in Tamale under the theme: "Reconciliation with God, Humanity and Nature in the Year of Mercy." In a communiqué, the bishops registered their displeasure about the growing incidence of land grabbing in the country and the indiscriminate acquisition of large tracts of land by multinational corporations, usually led by greedy and unpatriotic indigenes. They therefore advised that mercy must equally prompt our actions from harming our natural environment.[22] According to Rev. Fr. Wisdom Larweh, the communiqués are meant to appeal to the consciences of people, especially members of the Catholic church, and help them change their negative behavior toward the environment. He believes the words and opinions of bishops are respected and therefore taken seriously.[23]

It can be clearly deduced from the above discussions that the accumulated evidence of the human impact on the environment has prompted the Ghana Catholic Bishops' Conference to re-examine Scripture and existing social teachings, and to make stewardship ethics a component of Ghanaian

20. Ghana Catholic Bishops' Conference, "A Communiqué Issued by the Ghana Catholic Bishops' Conference at its Annual Plenary Assembly held at Sefwi Wiawso," (29th October—5th November, 2010), 1–5.

21. Ghana Catholic Bishops' Conference, "A Communiqué Issued by the Ghana Catholic Bishops' Conference at its Annual Plenary Assembly held in Sunyani," (8th—16th November, 2013), 5–6.

22. Ghana Catholic Bishops' Conference, "A Communiqué Issued by the Ghana Catholic Bishops' Conference at its Plenary Assembly held in Tamale" (October 7-14, 2016), 3–4.

23. Interview with Rev. Fr. Wisdom Larweh, The Assistant Secretary General of the Ghana Catholic Bishops' Conference, National Catholic Secretariat, Accra. March 2017.

Catholic theology. The concepts and language of sustainability interpreted by the bishops as the practical application of stewardship have become the overarching framework for environmental ethical duties in Catholic social teaching. This approach appeals to socially engaged Catholics because it builds organically on the themes of development and economic justice emphasized by Vatican II. Environmental initiatives deliberately borrow the sustainability framework from the United Nations linkage of economic justice and environmental protection from the environmental justice movement. For the bishops, these discourses provide a moral framework for enrolling others in their environmental initiatives.

The leaders of the Catholic church in Ghana have grafted environmental ethics onto human stewardship responsibilities to future generations. The Ghana Catholic Bishops' communiqués are important for conferring legitimacy on environmental concerns, but the most dynamic programmatic expression of the ethic has been the Abor Week celebrations, which bring Catholic social vision to bear on local issues through lay civil engagement. Collectively, this program constitutes the emergence of a practical theology of the environment for the Catholic church in Ghana. This program is relatively significant, yet is not well known due to the few resources allocated to it, but it portends an increase in public engagement with environmental concerns by the Catholic church in Ghana.

The Ghana Catholic Bishops' contribution to environmental ethics draws on Catholic social teaching and emphasizes continuity with its ethical vision, including social responsibilities for the economically marginalized and stewardship of the common good. These are not novel principles, but participants in these initiatives have expanded them to cover emerging moral concerns. Thus, Ghanaian Catholic bishops have drawn from their tradition and moral vision, reinterpreted key biblical texts, and have engaged laity in local discernment processes. By also drawing from contemporary environmental discourse about sustainability and justice, the Catholic church in Ghana is attempting to make its visions of stewardship practical for the Ghanaian culture. This is in line with Jürgen Moltmann's observation that even the command to subdue the earth pertains to extracting a vegetarian diet as in Genesis 1:29: "Then the Lord said I give you every seed-bearing plant on the face of the whole earth and every tree that has fruit with seed in it. They will be your food." Moltmann continues to explain, "Seeing that animals are also to exist on a vegetarian diet, the rule over animals means that human beings have the duty to peacefully co-exist with these animals."[24] Humans nonetheless are to treat all life with respect. The divine command is to be

---

24. Moltmann, *God in Creation*, 197.

understood as humans taking proper care of creation and not to subject creation to exploitation and destruction. The mandate implies living in harmony with the rest of creation. This rule must be extended beyond the selfish needs of humanity because care is to be given to every creature.

The Impact of Environmental Policies and Programs of the Catholic Church in Ghana on its Members

To ascertain the extent to which the policies, programs, and theologies of the local branches of the church reflect those promoted by the Ghana Catholic bishops, the study aimed to understand the degree to which the local branches are conversant with the church's environmental policies. A series of group discussions in eight congregations in the Greater Accra and Ashanti Regions are used to initiate a conversation about the relationship between the church and the environment. Groups were self-selected and averaged six participants in a group. Although we did not collect demographic data on the participants, anecdotal observation suggests that the majority of group participants were aged forty to sixty, with only two participants who appeared under forty years of age. The groups were held in rural, semi-urban, and urban contexts, and reflected a wide range of churchmanship.

Focus group discussions are used for two key reasons. First, focus-group discussions make room for intersubjective views, and second, it can serve a deeper function as a consciousness-raising group for future development on an issue. To reduce the possibility that the findings would inaccurately represent the population of Catholic church members, Morgan's suggestion is followed so that focus group facilitators take a more systematic approach to their organization.[25] To accomplish this, each of the discussion groups was given the same series of questions which took not more than one hour to complete. By using an interview guide to guide the focus group discussions, we were able to ensure that each session was conducted in a similar manner, thus enhancing the reliability of the findings.

Initial Findings:

The initial findings of the field work suggest that despite the effort made during various Ghana Catholic Bishops' conferences to engender a pro-environmental culture, there is very little evidence to suggest that individual church members can effectively connect their religious faith and

25. Morgan, "Focus Group Interviewing," 141–59.

their environmental dispositions. Very few people had heard of the church's policies and programs on the environment. With the exception of some members of the St. Peter's of Basilica Congregation (Roman Hill), Kumasi, all were surprised to hear about the Abor Week celebration and none of these congregations had ever taken part in celebrations.

One particular group participant announced that she had been involved in recycling campaigns and environmental activism within her community for her entire life and that she could see no connection between her own personal Christian faith (she is a lifelong Catholic and had served as an usher in the past) and her forty years of environmental service. When pressed further, she was unable to articulate any conscious relationship between her being a faithful Christian and her being an active environmentalist, and explicitly stated that the two aspects of her life were utterly unrelated to each other. Another participant from one of the discussion groups noted that despite the many interesting and informed views expressed by his environmentally minded group members, he was surprised that "none of us have made any attempt to link our Christian faith or study of the Bible with our ethical considerations of the environment and climate change." When pushed a bit further to reflect on this statement, he said, "I have never heard a message or sermon preached about stewardship or the environment in my years of regular church attendance, I wonder if priests even know that this (the environmental crisis) is a problem."[26]

Although the study had anticipated that an appreciable percentage of members of the local congregations of the church would be familiar with the exact environmental policies and programs of the church, a participant was surprised to find such a lack of theologically informed environmental knowledge within these particular groups. These groups were made up of individual attendees, and therefore they were expected to know more about the church's policies and programs. Moreover, most of these individuals were sufficiently concerned with the environmental issues in the country. This work originally was concerned that studying self-selected groups would unduly skew the data in a more pro-environment direction than would be representative of the diocese. Instead, this data, even with this possible bias, has no strong link between environmental concerns and the religious beliefs and practices of the respondents, as they relate to the church's official policies, programs, practices, and theologies.

---

26. Interview. Name withheld. April 2017.

View 1: Environmental Degradation as Sin

Respondents were asked to cast their minds on manufactured landscapes and discuss their experiences with industrialization and reflect on the human responsibility for environmental damage (pollution, waste disposal, deforestation, etc.). In these discussions, a consensus emerged from the groups that the profligate use of nature is tantamount to sin. All groups mentioned the problems related to overconsumption and the "throw it away syndrome," and all made some critical connections between the designs of disposable technologies, our desire to consume these technologies and our need to challenge both use and design.

It was interesting to observe within these discussion groups that the group of people who were cited as being both the most wasteful (and therefore the most sinful) and the most "green" (as indicated by their awareness of environmental problems and in recycling) were the youth. Indeed, when respondents spoke about damage to the creation, it was most frequently in the third-person plural: "Those young people today do not realize how connected we are to nature"; "they (youth) are so wasteful." The preference for framing wastefulness in the third-person plural indicated that the other rather than the self was the principal cause of environmental damage. Although it was unusual for respondents to claim personal or corporate responsibility for the current environmental crisis, one group member did suggest that there needed to be clearer teaching about the sinfulness of environmental destruction and as a result, there needed to be a forum for corporate repentance about it. In the context of the group, it was clear that repentance was both regarded as an act of contrition and a commitment to turning away from the sinful behavior. Although sin and waste were often conflated, this call for repentance was only mentioned once. Interestingly, the first person (singular or plural) was only used once in relationship to the causes of environmental damage.

View 2: Environment and Mission

Respondents saw the environment as a serious concern for the churches, but did not believe that the church was doing anything about it. On the history of the Catholic church's pro-environmental policies, programs, practices, and theologies, most of the respondents appeared both shocked that the church was so involved in the issue and surprised that they themselves were unaware of this involvement. When asked what they thought the church should do to address environmental issues, respondents called for clear teaching on the relationship between faith and the environment, specifically in the areas of waste disposal, reforestation, wise and ethical use of technology, sustainable living, and exposition on the gravity of environmental sins. They talked about the need for programs that could be organized in the local setting where the church's theological beliefs and ethical practices could be applied in the context of the community's corporate action—both as a sign of the church's love for God and creation and as a means of communicating the gospel in a relevant form to the world. Two groups made explicit mention of how the mission of the church (in the sense of proselytizing or evangelism) could be extended by providing a context for pro-environmental action for their communities.

It is suggested that the current environmental ethic that is being promoted by the Ghana Catholic bishops through the various communiqués (as reflected in the Abor Week campaign and the section on the environment in the various resolutions and pastoral letters) add other practical programs like safeguarding the creation. The safeguarding of creation ethic appears to convey an understanding of God as being immediately present within the ongoing task of the Catholic church's missional activity. The sources for this ethic come from a combination of the christological and liberation elements of the ecojustice ethic, with the emphasis on tending and managing the creation embodied by the stewardship ethic (what perhaps could be termed "Agro incarnational"—the church's desire to tend the earth in order to support the poor and marginalized). When examining the root of the present environmental crisis, the safeguarding creation ethic points to individual consumptive practices which must be addressed within the church community, as well as broader corporate practices that are endemic to the whole of society.

Elsewhere, Laurel Kearns describes a creation spirituality ethic, which she finds represented in the margins of liberal ecumenical Christian churches. Unlike the ethics noted above, it neither draws its resources from biblical agrarian imagery (stewardship) nor from christological concerns reflected in liberation theology (ecojustice). Instead, creation spirituality reflects trends in new age spiritualism and deep ecology, and prefers a pantheistic rather than transcendent God whose relationship with creation is principally characterized by radical immanence.[27] For this view, the root of the environmental problem is anthropocentrism and the resultant human alienation from nature; consequently, addressing the environmental crisis will require the widespread adoption of a holistic spirituality and a new way of seeing the world. Thus, what is less prominent in Catholic

---

27. Kearns, "Saving the Creation," 55–70.

pro-environmentalism are those themes associated with Kearns's typology of the creation spirituality ethic, although the turn away from anthropocentrism and toward a more inclusive understanding of the creation is consistent with much of the safeguarding of the creation ethic materials. Similarly, whereas this promotes a cure for the environmental crisis, whereas it is the individual who is responsible for acting as a faithful steward within other environment programs like Kearns's characterization of the stewardship ethic,[28] and whereas it is the collective that is responsible for promoting the social justice work of the church in the ecojustice ethic, the safeguarding of the creation ethic requires action from individuals, their congregation, diocese, and the institutional church.

A Way Forward

Given the findings of the study and their implications, it is necessary to come out with some recommendations that may help to salvage our environment. Therefore, the following recommendations are made.

First, the theologians and the church should act as a forum for various professionals and experts to meet and discuss environmental issues to bring their expertise to bear on the problems confronting us. Some University dons and other professionals, who are experts in various disciplines, are members of the church. Some only go to church on Sundays to warm the pews without realizing that they can use their store of knowledge to help humanity and make this world a better place. The church should create the enabling environment for these professionals to provide expert advice on various issues. Such issues that may be discussed include the following: waste disposal, galamsey, chain-saw operations, sustainable and ecological farming practices, early warning systems, system management, rural investment to reduce the long-term effects of climate variability on food security, and better agricultural and land use practices.

Second, reports from the church discussions could be used as a starting point for advocacy. Theologians and the church should be a strong advocacy group to check on both the government and private companies whose operations have a direct impact on the environment to see how they are mitigating the effects their operation is having on the environment. Activities of logging companies who export significant quantities of wood should be made to pay attention to replenishing what they take out. The government should be lobbied to make laws that will help in the conservation and preservation of the environment. The church should also

28. Kearns, "Saving the Creation," 60.

help to monitor the enforcement of these laws and call the attention of the government to lapses, whilst suggesting strategies that will help increase the implementation of these laws. In doing so, the church will be playing its role as the prophetic voice of our day, pointing out to the people of Ghana the consequences of overexploitation of the land and reminding Ghanaians of the creation mandate. The message to the world should include humanity discovering its rightful place in the universe, and its role of helping the entire universe to function in a manner that brings honor and glory to its Creator. It needs to be emphasized that all things exist for the glory of God, and humans are accountable to God for their use of the earth. Their commission is to rule nature as God's representative, treating nature as the God who created it would.

Finally, theologians and the church must endeavor to develop a relevant Ghanaian (African) Christian theology and policy for the environment that will guide the church in the quest to contribute to environmental sustainability efforts in Ghana. Such a theology and environmental policy must integrate relevant traditional African indigenous beliefs and environmental usages that will be relevant to the Ghanaian Christian context. The Ghanaian Christian approach to environmental ethics can inspire praxis, responsiveness to the relationship, and a deep commitment to care. The church must cooperate with traditional leaders in their quest for environmental sustainability efforts in Ghana. Besides, preaching and teaching on environmental sustainability should become a regular practice in every local church and preaching station or post so that we can fulfill a holistic ministry of human as well as cosmic redemption and win everything for Christ.

One suspects that one of the weaknesses of the top-down approach taken by the Ghana Catholic Bishops' conference has been its reliance on a highly didactic mode of pro-environmental education. Paul Djupe and Patrick Kieren Hunt, in their analysis of the disparity between the pro-environmental view of clergy in the Episcopal and Lutheran churches and the more conservative views regarding the environment held by their congregations, ask whether clergy should communicate with the goal of reshaping beliefs about the meaning of religious text or speak directly about the nature of environmental problems.[29] Djupe and Hunt suggest that proclamation and education are not the most effective ways for a cultural ideal to impact the actions and attitudes of individuals. Apart from education, Djupe and Hunt believe that to change behaviors, the churches need to employ subtler and more diffuse methods of communication; methods that create a vast communication network through which theological, ethical,

29. Djupe and Hunt, "Beyond the Lynn White Thesis," 670–86.

and political knowledge are transmitted. That is not to suggest that theologians and the church in Ghana can simply rely on emerging forms of social networking technology to communicate its environmental policy. Rather, a more intentional and creative way of bringing together faith environmentalism seems to be necessary.

Conclusion

This study indicates that the church, although informed by numerous valuable modes of local wisdom and pastoral concerns, has failed to successfully identify with the bigger context of environmental action. What is required, it would seem, is a more effective means of communicating the church's environmental policies, practices, and theologies with the local branches (churches). Despite a significant campaign to heighten the church's awareness of environmental issues, and motivation from the government to mobilize faith communities into pro-environmental action, the local church and the institutional church seem ill equipped to link their theological self-understanding with some of their congregants' nascent pro-environmental orientations. To be sure, the institutional church's expansive desire to model environmental practice is a laudable aim that may well help to place the church in what Roger Gottlieb calls the "green public sphere," a common space shared by religious and nonreligious individuals alike, for whom environmental concern is a public issue that should be discussed and acted upon in the open. As one of many other pro-environmental agents, the church (institutional and local) contributes to the green public sphere by helping to normalize pro-environmental discourse at all levels of society and to support a collective culture of environmental awareness and action to foster the capacity for ecological self-examination by individuals, communities, and nations.[30]

While the Ghana Catholic Bishops' conferences are clearly attempting to contribute to the green public sphere, the effectiveness of their contributions has been limited in the local context. Although politicians and activists may be well versed in the church's activities, the local context could benefit from more dynamic, creative, and intentional efforts so that those in the church with a preexisting interest in environmental matters could become aware of the resources of their own faith communities.

Beyond this study, Catholic environmental policies, programs, practices, and theologies are as diverse as Catholicism itself. There are clearly locations where environmental theology has greater resonance than others, and clearly there are dioceses in which environmental concerns are

30. Gottlieb, *Greener Faith*, 236.

not regarded as pressing missional or indeed practical issues. What this study does indicate, however, is that there is a significant communication problem within the church, a problem that intuition and anecdotal observations seem to suggest extends well outside of Catholicism and well outside of environmental beliefs and practices. The disparity between the beliefs and practices on the ground and the theologies and imperatives at the top surely represent a crucial problem that impinges upon any church's ability to express its newly developed environmental concern. Thus, it is time for theologians in Ghana to rethink how to raise awareness about the development of the whole creation, thereby adopting a more concerted strategic approach. The proposals in this presentation should motivate a shift from merely issuing of resolutions, messages, and communiqués to a practical strategic approach that will integrate all responsible stakeholders, especially the members of the church's local branches. For emphasis, theologians' involvement in salvaging the environment should be a priority, for this is indeed a security threat to all the diverse ways of human development. The church should not only cling to the issuing of communiqués, resolutions, and messages but more importantly, should focus on practical implementation. Let us remember that environmental degradation affects every facet of life, including our very survival, our religion, and our worship. Theologians therefore must reorient themselves to adequately improve upon their efforts at environmental sustainability because the church's mission involves both humanity's and nature's redemption.

Bibliography

Adegoke, Jimmy. *African's Environmental Challenges in the 21st Century: Current and Emerging Issues*. Accra, Ghana: Last Word, 2016.

Adow Obeng, Emmanuel, and Mary N. Getui, eds. *Theology of Reconstruction*. Nairobi: Acton, 2003.

Anderson, Kerby, *Christian Ethics in Plain Language*. Nashville: Thomas Nelson, 2015.

Asante, Emmanuel. "Ecological Crisis: A Christian Answer." *Trinity Journal of Church and Theology* 4 (December 1994) 1–9.

Collins, Paul. *God's Earth: Religion as if Matter Really Mattered*. Dublin: Gill and MacMillan, 1995.

Djupe, Paul A., and Patrick Kieren Hunt. "Beyond the Lynn White Thesis: Congregational Effects of Environmental Concern." *Journal for the Scientific Study of Religion* 48.40 (2009) 670–86.

Ghana Catholic Bishops' Conference. "A Communiqué Issued by the Ghana Catholic Bishops' Conference at its Annual Plenary Assembly held in Sunyani," (November 8–16, 2013), 1–16.

Ghana Catholic Bishops' Conference. "A Communiqué Issued by the Ghana Catholic Bishops' Conference at its Plenary Assembly held in Tamale," (October 7–14, 2016), 1–23.

Golo, Ben-Willie Kwaku. "Christian Thought and the African Experience: A Survey of Approaches to Theology in the Contemporary African Academy." In *Unpacking the Sense of the Sacred: A Reader in the Study of Religions, Arts and Humanities Series No. 7*, edited by Abamfo Atiemo et al., 68–89. Accra, Ghana: Ayebia Clarke, 2014.

Gottlieb, Roger S. *A Greener Faith: Religious Environmentalism and Our Planet's Future* Oxford: Oxford University Press, 2006.

Kearns, Laurel. "Saving the Creation: Christian Environmentalism in the United States." *Sociology of Religion* 57 (1996) 55–70.

Mante, Joseph O. Y., *Africa: Theological and Philosophical Roots of Our Ecological Crisis*. Accra, Ghana: SonLife, 2004.

Moltmann, Jürgen. *God in Creation*. London: SCM, 1985.

Morgan, David. "Focus Group Interviewing." In *Handbook of Interview Research: Context and Method*, edited by Jaber Gubrium and James Holstein, 141–59. Thousand Oaks, CA: Sage, 2002.

Pope John Paul II. "Papal Message on the Celebration of the World Day of Peace, January, 1, 1992." http://w2.vatican.va/content/john-paul-ii/en/messages/peace/documents/hf_jp-ii_mes_20031216_xxxvii-world-day-for-peace.html.

Raphael, Noah A. *Abor Week: A Response to Environmental Care*. Takoradi, Ghana: St. Francis, 1991.

Santimire, H. Paul. *The Travail of Nature: The Ambiguous Ecological Promise of Christian Theology*. Minneapolis: Fortress, 1985.

Squicciarini, Donato. *Pope John Paul II's Message for the World Day of Peace*. New York: Duncker & Humblot, 2005.

The Ghana Catholic Bishops' Conference. *Ghana Bishops Speak: A Collection of Communiqués, Memoranda and Pastoral Letters of the Ghana Catholic Bishops' Conference*. Accra, Ghana: National Catholic Secretariat, 1999.

The Ghana Catholic Bishops' Conference. *Ghana Bishops Speak: A Collection of Communiqués, Memoranda and Pastoral Letters of the Ghana Catholic Bishops' Conference*. Accra, Ghana: National Catholic Secretariat, 2007.

The Ghana Catholic Bishops' Conference. *Ghana Bishops Speak: A Collection of Communiqués, Memoranda and Pastoral Letters of the Ghana Catholic Bishops' Conference*. Accra, Ghana: National Catholic Secretariat 2008.

The Ghana Catholic Bishops' Conference. *Ghana Bishops Speak: A Collection of Communiqués, Memoranda and Pastoral Letters of the Ghana Catholic Bishops' Conference*. Accra, Ghana: National Catholic Secretariat, 2010.

Part III

Impact of Artisanal and Small-scale Mining on the Environment in Ghana

ADWOA BOADUA YIRENKYI-FIANKO

Abstract

This paper highlights the problems and impact of Artisanal and Small-scale mining on the land, water bodies, and economy in Ghana. Mining has made a positive contribution to the economy of Ghana through significant earnings accrued from the export of gold and other minerals. Despite these benefits, mining, especially Artisanal and Small-scale Mining (ASM) activities have had a negative impact on the environment through the pollution of waterbodies and land, as well as the economic growth of the country. The phenomenon of illegal mining, popularly referred to as galamsey, is generally unsafe and socially and environmentally destructive. Although Ghana has several environmental protection laws and policies, their implementation has not been able to address the environmental challenges caused by ASM activities. The call is for civil society and the church to add their voice and actions to complement what the state is doing to salvage the environment.

Introduction

THIS PAPER EXPLORES THE negative impact of Artisanal and Small-scale Mining (ASM) in Ghana. It looks at the impact on waterbodies, the land, and the economy as factors that necessitate concern about the environment. Whatever happens to the environment affects the people. ASM is generally accepted as a practice that involves the use of rudimentary processes and equipment and requires minimal capital investment. It involves basic techniques of mineral extraction, highly manual processes, hazardous working conditions, and frequently negative human and environmental health impacts.[1]

In recent times, there have been several deliberations on various media platforms in Ghana with regards to the high prevalence of ASM operations, especially illegal mining, popularly and locally referred to as *Galamsey*, which has confounded the minds of many Ghanaians. These include licensed (legal) small-scale mining operators and unlicensed (illegal) small-scale mining operators (those who operate without permission). Both licensed and illegal artisanal and small-scale miners tend to use the same unacceptable mining methods. Hilson and Potter observe in their study that there is little difference, either organizationally or technologically, between legal and illegal mining activities, apart from the fact that the licensed mining activities have security of tenure on a demarcated mineralized concession for a given period of time.[2] Unfortunately, the activities of both the legal and illegal mining operations have had negative impacts on the environment, although the illegal ones who are not regulated tend to cause more harm. Over the years, the Government of Ghana, together with other agencies, has implemented policies and put in place various measures to control and curb these activities that are destroying the environment, but these efforts have not been very successful.[3] What are some of the effects of such abuses on land, water resources, human health, the health of other living organisms, and the economy? It needs to be noted that large-scale mining activities also contribute to environmental degradation, but the attention here is on artisanal and small-scale ones.

A View on ASM in Ghana

ASM operations began in Ghana in the sixth century when it was discovered that there were traces of precious minerals such as gold in alluvial deposits. For many decades, the activities of artisanal and small-scale miners were not formally recognized and monitored until the implementation of the Economic Recovery Programme (ERP) in the 1980s. Ghana has long been known for its gold reserves and was therefore given the name The Gold Coast during its colonial era.

---

1. Hilson, "'Fair Trade Gold,'" 386–400.
2. Hilson and Potter, "Why Is Illegal Gold Mining?" 237–70.
3. Aryee et al., "Trends in the Small-scale Mining," 131–40; Hilson et al., "Improving Awareness of Mercury Pollution," 275–87.

ASM has dominated the mining industry in Ghana from traditional times through the colonial period and the early independence period to the present era.[4] It is significant to note that small-scale mining activities were abolished during the colonial era when the Europeans introduced large-scale gold mining.[5] Yet, small-scale gold mining has contributed and contributes, to the production of gold in Ghana and the creation of employment for the unskilled labor force in rural communities after the ban on small-scale gold mining was lifted in 1989 (when the Small-Scale Mining Law, 1989 was passed). Many small-scale miners use basic equipment such as dredging boats, water pumps, pickaxes, shovels, chemicals, and in recent times, excavators. Since most of the operators are unskilled in working with unplanned measures with no checks and plans for reclamation, ASM has had negative effects on the environment. It is common knowledge that the miners seek for gold and other mineral resources, but do not give the environment the attention it deserves.[6] They do not care what happens to the pits they dig, how the dirt and chemicals pollute the rivers and waters they use to wash the mineral, and what becomes of the land they have exploited.

Impact of ASM on the Land and Health

ASM activities have negatively impacted the land in a number of countries, including Ghana.[7] ASM activities have led to discharge and run-off of mine waste into rivers, ponds, streams, wells, and boreholes for drinking water, and has resulted in severe heavy metal contamination, and deforestation.[8] The levels of heavy metals such as arsenic, lead, and cadmium have had, and continue to have, serious effects on the land.[9] People who live near sites where these metals have been improperly disposed of are exposed to these metals by ingestion through indirect drinking and eating or inhalation.[10] According to Anderson, ASM is "characterized by a vicious cycle of discovery, migration, and relative economic prosperity followed by resource depletion, out-migration and economic destitution and after the depletion of the reserves, sites are abandoned, and the community is left to cope with a legacy

---

4. Sarpong, "Sweat and Blood," 346–62.
5. Kessey and Arko, "Small Scale Gold Mining," 12–30.
6. Sarpong, "Sweat and Blood," 346–62.
7. Hilson et al., "Improving Awareness of Mercury Pollution."
8. Bortey-Sam et al., "Health Risk Assessment," 1–12.
9. Obiri et al., "Human Health Risk Assessment."
10. Martin and Griswold, "Human Health Effects."

of environmental devastation and extreme poverty."[11] Unfortunately, many people fail to realize that humanity's existence is totally dependent on the land and environment, and thus all human engagements are dependent on nature's services.[12] Some heavy metals released through mining activities of ASM are deposited and accumulated in environmental sinks, such as water, and are absorbed into sediment surfaces. In fact, some of these heavy metals are dangerous to living organisms even in very small quantities. Arsenic, cadmium, manganese, mercury, lead, and other metals have exceeded guideline values in water and soil samples collected from some mining communities and samples from some ASM sites have exceeded guidelines for acidity, turbidity, and nitrates.[13] Arsenic contamination of groundwater has been reported in a number of countries, including Ghana, and prolonged drinking of arsenic-contaminated water can result in arsenicosis, which can lead to slow and painful death.[14] Mercury pollution has already been identified as a lingering problem in several of Ghana's important small-scale gold mining communities,[15] and this further leads to the contamination of drinking water sources. According to Siegel, different water quality levels are required for various uses and failure to meet the water quality standards for these specific uses can result in sickness in humans and even death.[16] Removal of vegetation cover, soil erosion, and siltation, and the creation of open excavations are among the many deathtraps associated with ASM.

Impact of ASM on Waterbodies

Water is an indispensable natural resource that is necessary for the existence of living organisms in order for them to thrive successfully in their habitat. There is therefore a need for water resources to be protected and preserved. Ghana is well endowed with water resources, and Sarpong suggests there is an estimated total actual renewable water sources of 53.2 billion m^3 per year.[17] Ghana has five basins: the Densu River basin, Ankobra Basin, Pra Basin, Tano Basin, and White Volta Basin. The Ministry of Water Resources, Works and Housing has the overall responsibility for water management

---

11. Anderson, "Phytoextraction to Promote Sustainable Development," 51–56
12. Hill, *Understanding Environmental Pollution*.
13. See Bortey-Sam et al., "Health Risk Assessment," 1–12; Tetteh et al., "Levels of Mercury," 635–43.
14. Ahuja, *Monitoring Water Quality*.
15. Hilson et al., "Improving Awareness of Mercury Pollution."
16. Siegel, *Demands of Expanding Populations*.
17. Sarpong, "Sweat and Blood," 346–62.

and supply in Ghana. Other governmental institutions involved in potable water production, management, and distribution are the Water Resources Commission (which is involved in water resources management in the country), the Water Directorate, the Environmental Protection Agency, the Ghana Water Company Limited, the Community Water and Sanitation Agency, and the Public Utilities Regulatory Commission.[18]

In Ghana, water resources are mainly used for human consumption, farming, and irrigation.[19] Domestic and industrial urban water supplies are based almost entirely on surface-water resources like rivers and streams, which are believed to be sufficient to meet present and future consumptive water demands. In spite of the availability of water to meet present and future demands, there is a shortfall in water distribution and a national demographic and household survey found that only 40 percent of urban residents have treated water in their homes.[20] Unfortunately, the quantity and quality of water in river bodies, which serve as the main source of water for drinking, household chores, and other activities, especially for rural communities, are being destroyed by small-scale mining activities.[21] ASM largely depends on water to process what is dug before it is refined.

Small-scale mining operations mainly mine alluvial ore minerals deposited along the bank and/or bed of waterbodies, such as streams and rivers, to obtain minerals such as gold, diamond, bauxite, etc. Artisanal and small-scale miners, with their low-level technology, can only extract gold and other resources in its free form and not in its compound form. Therefore, the alluvial deposits of free gold are easier to mine on land surfaces than in rocks, and that is why most artisanal and small-scale miners either mine in river beds or along rivers,[22] and end up polluting water bodies. ASM sites require water for a variety of functions such as sluicing or washing, panning, and amalgam preparation, and therefore mining activities need to be located near water sources.[23] Where waterbodies are not close by, they create means and channels for rivers and streams to the mining sites.

Mine tailings, which are mostly crushed ore and rock from mining operations, pose potential threats to water quality and human health. Unfortunately, proper management of tailings and waste often requires transporting them to offsite locations, which is costly. The mine tailings are therefore

18. Sarpong, "Sweat and Blood," 346–62.
19. Owusu et al., "Review of Ghana's Water Resource."
20. Sarpong, "Sweat and Blood," 346–62.
21. Kessey, and Arko, "Small Scale Gold Mining."
22. Armah et al.. "Assessing Environmental Exposure," 786–98.
23. Rajaee, et al., "Integrated Assessment," 8971–9011.

generally allowed to sit indefinitely in communities, either in sedimentation ponds or piles.[24] The concentrations of heavy metals such as arsenic, lead, mercury, and cadmium, which naturally pollute the environment, have increased in artisanal mining communities.[25] A study conducted in Ghana in 2010 indicated that for licensed small-scale gold mining operators in the Offin River, only 21 percent treat their tailings before discharging them into the Offin River, 52 percent discharge directly into the river without treatment, and 27 percent store them in mining pits.[26] It is common knowledge that most of the waterbodies and rivers in Ghana have been discolored with the discharges, causing all living organisms and fishes in the waterbodies to die. Such waters, which were useful for human consumption, animal consumption, and agricultural purposes, are no longer suitable.

According to the Executive Secretary of the Water Resources Commission, Ghana, about 60 percent of Ghana's water bodies have been polluted as at 2017, with many in a very critical condition, and the polluted water bodies are mostly in the south-western parts of the country, where ASM operations are widespread.[27] Due to a shortage of affordable potable water, the UN estimates that, globally, about 1.2 billion people are forced to drink unclean water, which leads to water-related diseases that kill about 5 million people each year, mostly children. The UN also estimates that 2.7 billion people will face water shortages by 2025.[28] The Nuclear Chemistry and Environmental Research Centre (NCERC) of the Ghana Atomic Energy Commission (GAEC) has warned that Ghana's groundwater reserves risk being contaminated if ASM activities, especially the illegal ones that are often unsafe and environmentally destructive, are not properly regulated.[29] Water plays an indispensable part of our daily lives and there is therefore the need for the quantity and quality of waterbodies to be preserved through the implementation and enforcement of policies and regulations, and the treatment of the already-polluted water.

A study conducted in the Tarkwa area in southwest Ghana, found the pH to be 5.71 for boreholes, 5.05 for wells, and 6.40 for stream water, which are all below the World Health Organization (WHO) guideline of 6.5–8.5. The study observed that samples from mining areas had very low pH, all the samples were above the WHO guidelines for turbidity, and 80

24. Akabzaa and Darimani, *Impact of Mining Sector Investment*.
25. Obiri et al., "Human Health Risk Assessment."
26. Kessey and Arko, "Small Scale Gold Mining."
27. Water Resources Commission report, 2017.
28. Ahuja, *Monitoring Water Quality*, 14.
29. Ghana Atomic Energy Commission, "Illegal Mining Activities."

percent of all samples did not comply with the WHO guideline for chemical oxygen demand.[30]

Impact of ASM on the Economy

Ghana is a well-known producer of gold, diamond, manganese, and bauxite, among others in Sub-Sahara Africa. Ghana is the second-largest producer of gold in Africa. The Ghanaian economy as a whole, as well as many rural communities, have benefitted from small-scale mining activities through the earnings accrued from the sale and export of gold. ASM also provides employment for a number of rural folks and unskilled laborers. The mining sector plays a significant role in building the economic earnings of the country in terms of foreign exchange earnings, employment generation, mineral royalties, employee income, taxes, payments, etc. In fact, the nation's GDP increased from 1.3 percent in 1991 to an average of about 5.2 percent between the years 2001–2004, with gross foreign exchange earnings also increasing progressively from 15.6 percent in 1986 to 46 percent in 1998, representing US$ 124.4 million in 1986, and US$ 793 million in 1998 from the proceeds of mining. Between 2004 and 2005, it increased from US$798 million dollars to US$995.2 million dollars. In 2006, the mining sector accounted for approximately 40 percent of Gross Foreign Exchange (GFE) earnings, which was approximately 5.2 percent of GDP.[31] The proceeds from ASM cannot be left out of these achievements.

ASM operations mainly employ unskilled labor and this leads to a reduction in the rate of unemployment among the youth within rural mining communities, and thus a reduction in armed robbery cases and other social vices. Unfortunately, despite these benefits, ASM activities especially *galamsey*, have had devastating effects on the country in the form of pollution of water bodies, destruction of farm lands, loss of human lives, and increased high school dropout rate, amongst others, which invariably affects the economy. ASM operations exist in a number of countries around the world, including Ghana, and these activities are common in most mining communities in Ghana. The ASM industry has contributed to the economic growth of Ghana, a developing country in West Africa, although its activities have had adverse effects on the environment.[32] The nation needs a lot of money to correct the harm caused by ASM.

30. Armah et al., "Assessing Environmental Exposure," 736–98.

31. Ghana Minerals Commission, *Statistical Overview of Ghana's Mining Industry*, 10.

32. Rajaee et al., "Integrated Assessment," 8971–9011.

In 2016, the Ministry of Lands and Natural Resources in Ghana reported that illegal mining cost the country $2.3 billion in lost revenue. Likewise, the production of cocoa, which is the country's third-largest export, is being threatened by these ASM activities because cocoa farms are being destroyed by these miners for precious minerals. They take over farms that have gold deposits, destroy the crops, and mine the gold. A number of farmers in mining communities have sold their farms to miners for mineral exploration for meager amounts of money, thus causing the nation to lose revenue from cocoa exports.

The activities of ASM, as earlier indicated, also affect the health of the people. A number of mining communities are filled with open mine pits and these pits are subsequently filled with storm water and thus serve as breeding places for mosquitoes. Children and livestock fall into open mine pits on a regular basis, never to return. News of deaths resulting from collapsed pits are reported regularly, but unfortunately the death toll has not served as a deterrent to other miners who later enter the sites.

The Way Forward

It is said that before the United Nations Conference on Human Environment held in Stockholm, Sweden in 1972, environmental protection did not feature extensively in The Minerals Ordinance (CAP 185) and Minerals Act, 1962 (Act 126), the main laws regulating mining in the country then. With increasing awareness of the environmental problems, laws and policies were passed. The Legislative Instrument, 1970 (LI 665) was amended by LI 689. Other laws passed include Minerals and Mining Act, 2006 (Act 703), Environmental Protection Agency Act, 1996 (Act 490), and Minerals Commission Act, 1993 (Act 450), etc.[33]

Although a number of laws and policies have been put in place by the Government of Ghana to curb the negative impacts of ASM on the environment, the bold step to ban all artisanal and small-scale mining in 2017 seems to be a great intervention. It started with a temporary ban for six months beginning February 1, 2017. However, the ban was extended for another three months and later made indefinite. In the process, the government of Ghana begun discussions on the way forward. It came out with a Multi-sectorial Integrated Mining Project (MMIP) and alternative livelihood interventions. It later began training about 3,000 licensed artisanal small-scale miners on ethical mining at the George Grant University of Mines and Technology, Tarkwa (GGUMaT). In addition, fifteen media

33. Kessey and Arko, "Small Scale Gold Mining," 15–16.

practitioners received training on sustainable mining at the GGUMaT. The government also engaged stakeholders and organized courses for chiefs, public sector workers, operators in the national security, religious leaders, the National Youth Council, etc.[34] Eventually the ban was lifted in October 2018, and ASM operations resumed.

It needs to be emphasized that civil society and the church also have a great role in complimenting the efforts of the government. Although religious leaders were included in the public education, it seems the Christian church is still not well equipped to join in the campaign. The onus lies on the religious leaders, who were trained to consciously reach out to the other leaders and members.

Despite its negative impacts, ASM plays an essential role in developing societies, for it serves as a major source of income for some rural communities. There is, therefore, a need for a sustainable approach to ASM to be adopted to protect the environment. The state and church must not only think about their gains but how people earn their money. Income from ASM must not only enrich pockets, but also be partly used to develop communities. It is important for ecotheologians to understand the scale, scope, and operations of ASM (both licensed and unlicensed), and the attitude of the mine workers and inhabitants of the mining communities.[35]

Solving the problem of environmental degradation will also require policies that will cater to the welfare of low-income earners in the country, awareness creation of the impact on the environment, training of miners on environmental sustainability issues, and a fight against corruption. It is important for local community dynamics to be assessed in order to adopt community-based approaches for illegal mining education, training, and technological assistance.[36]

Finally, the lack of infrastructural development and social facilities in the communities whose land are used for mining are enough evidence that revenue from royalties have not been properly managed to meet the needs of the people. Civil society and the church must call on the central government, metropolitan, municipal and district assemblies and the Office of the Stool Lands Administration to be accountable, and do more for the ordinary citizen to benefit from the royalties.

---

34. Government of Ghana, "Government Presents Road Map."
35. Tschakert, "Recognizing and Nurturing Artisanal Mining," 24–31.
36. Hilson et al., "Improving Awareness of Mercury Pollution."

Conclusion

This paper has traced the activities of ASM in Ghana, and discussed its impact on the land, waterbodies, health, and the economy. It cannot be overemphasized that mining has made a positive contribution to the economy of Ghana, yet has negatively impacted the environment. It was argued that the interventions made to bring sanity to ASM have not been holistic. The lifting of the ban on ASM by the government in 2017, and subsequently putting measures in place in 2018, should mean that civil society and the church, as the conscience of the society, will actively make their voices heard going forward. Some accruals and economic gains from ASM must be used to develop communities. There must be education of the people of the local communities so they can engage in ethical approaches in mining, and accountability from those who receive the royalties must be seen.

Bibliography

Ahuja, Satinder. *Monitoring Water Quality: Pollution Assessment, Analysis and Remediation.* Amsterdam: Elsevier, 2013.

Akabzaa, Thomas, and Abdulai Darimani. *Impact of Mining Sector Investment in Ghana: A Study of the Tarkwa Mining Region.* Washington, DC: SAPRIN, 2001.

Amankwah, Richard K., and Carl Anim-Sackey. "Strategies for Sustainable Development of the Small-scale Gold and Diamond Mining Industry of Ghana." *Resources Policy* 29.3–4 (2003) 131–38.

Anderson, Chris W. N. "Phytoextraction to Promote Sustainable Development." *Journal of Degraded and Mining Lands Management* 1.1 (2013) 51–56.

Armah, Frederick A., et al. "Assessing Environmental Exposure and Health Impacts of Gold Mining in Ghana." *Toxicological & Environmental Chemistry* 94.4 (2012) 786–98.

Aryee, Benjamin N. A., et al. "Trends in the Small-scale Mining of Precious Minerals in Ghana: A Perspective on its Environmental Impact." *Journal of Cleaner Production* 11.2 (2003) 131–40.

Bempah, Charles K., and Anthony Ewusi. "Heavy Metals Contamination and Human Health Risk Assessment around Obuasi Gold Mine in Ghana." *Environmental Monitoring and Assessment* 188.5 (2016) 261–72.

Bortey-Sam, Nesta, et al. "Health Risk Assessment of Heavy Metals and Metalloid in Drinking Water from Communities Near Gold Mines in Tarkwa, Ghana." *Environmental Monitoring and Assessment* 187.7 (2015) 1–12.

Geenen, Sara. "A Dangerous Bet: The Challenges of Formalizing Artisanal Mining in the Democratic Republic of Congo." *Resources Policy* 37.3 (2012) 322–30.

Ghana Atomic Energy Commision "Illegal Mining Activities." https://gaecgh.org/en/ghanas-groundwater-reserves-threatened-by-illegal-mining-activities/.

Hill, Marquita. *Understanding Environmental Pollution.* Cambridge: Cambridge University Press, 2010.

Hilson, Gavin. "'Fair Trade Gold': Antecedents, Prospects and Challenges." *Geoforum* 39.1 (2008) 386–400.
Hilson, Gavin, and Clive Potter. "Why Is Illegal Gold Mining Activity So Ubiquitous in Rural Ghana?" *African Development Review* 15.2–3 (2003) 237–70.
Hilson, Gavin, et al. "Improving Awareness of Mercury Pollution in Small-scale Gold Mining Communities: Challenges and Ways Forward in rural Ghana." *Environmental Research* 103.2 (2007) 275–87. https://doi.org/10.1016/j.envres.2006.09.010.
Kessey, Kwako D., and Ben Arko. "Small Scale Gold Mining and Environmental Degradation in Ghana: Issues of Mining Policy Implementation and Challenges." *Journal of Social Sciences* 5.1 (2013) 12–30.
Martin, Sabine, and Wendy Griswold. "Human Health Effects of Heavy Metals. Environmental Science and Technology Briefs for Citizens." *Center for Hazardous Substance Research* 15 (2009) 1–6.
Obiri, Samuel, et al. "Human Health Risk Assessment of Artisanal Miners Exposed to Toxic Chemicals in Water and Sediments in the Prestea-Huni Valley District of Ghana." *International Journal of Environmental Research and Public Health* 13.1 (2016) 139–49.
Owusu, Phebe A., et al. "A Review of Ghana's Water Resource Management and the Future Prospect." *Journal of Cogent Engineering* 3.1 (2016) 73–88.
Rajaee, Mozhgon, et al. "Integrated Assessment of Artisanal and Small-Scale Gold Mining in Ghana—Part 2: Natural Sciences Review." *International Journal of Environmental Research and Public Health* 12.8 (2015) 8971–9011.
Sarpong, George A. "Customary Water Laws and Practices: Ghana." www.foa.org/fileadmin/templates/legal/docs/CaseStudy_Ghana.pdf.
Sarpong, Samuel. "Sweat and Blood: Deific Interventions in Small-Scale Mining in Ghana." *Journal of Asian and African Studies* 52.3 (2017) 346–62.
Sidibé, Yoro. "Flood Recession Agriculture for Food Security in Northern Ghana: Literature Review on Extent, Challenges, and Opportunities." https://www.researchgate.net/Hydrological-map-of-Ghana_fig2_304592560.
Siegel, Frederic R. *Demands of Expanding Populations and Development Planning: Clean Air, Safe Water, Fertile Soils*. Berlin: Springer, 2008.
Tetteh, Samuel, et al. "Levels of Mercury, Cadmium, and Zinc in the Topsoil of Some Selected Towns in the Wassa West District of the Western Region of Ghana." *Soil and Sediment Contamination* 19.6 (2010) 635–43.
Tschakert, Petra. "Recognizing and Nurturing Artisanal Mining as a Viable Livelihood." *Resources Policy* 34.1–2 (2009) 24–31.
Walser, Georg. "Economic Impact of World Mining." *International Atomic Energy Agency* 315 (2000) 263–64.

Exploring Possible Solutions to the Galamsey Menace in Ghana: The Christian Church in Perspective

John Appiah and James Kwaku Agyen

Valley View University, Ghana

Abstract

The galamsey *menace in Ghana has, in recent times, attracted the attention of opinion leaders and the ordinary Ghanaian, as well as the outside world. Many suggestions have been made to curtail the daring effects of* galamsey *in Ghana. This article proposes how the Christian church can play a mediatory role in dealing with the* galamsey *menace in Ghana using a biblical-ecological approach. It argues that the church must explore ways of affecting the hearts of the people on divine ownership and human stewardship of land and other natural resources to make a real impact.*

Introduction

OVEREXPLOITATION OF MINING LAND in Ghana has, of late, attracted the attention of the public, and key stakeholders of the mining industry in particular.[1] The unchecked activities of small-scale miners have destroyed the forest, caused the polluting of river bodies, and caused destruction of farmlands, among others, in areas of their operations and beyond. These small-scale miners include Ghanaian and non-Ghanaian nationals. The

---

1. Mantey et al., "Operational Dynamics of 'Galamsey,'" 11.

government of Ghana in 2017 placed a temporary ban on small-scale mining, popularly known as *galamsey* in the country. The government has also instituted a joint police and armed force, called "Operation Vanguard," to enforce compliance of the temporary ban of *galamsey* operations in Ghana. The temporary ban on small-scale mining has attracted the displeasure of the Association of Small-scale Miners in Ghana and a section of the Ghanaian populace. Some are calling on the government of Ghana to lift the temporary ban on small-scale mining to enable Ghanaians to resume their operations. They argue that the ban has led to financial burdens on small-scale miners, some of whom have to settle bank loans they contracted for their services. The ban has also led to unemployment of small-scare mining workers, resulting in untold economic hardships to the miners and their families, to say nothing of the communities that thrive on the income of the workers. Some members of the Association of Small-scale Miners are accusing the government of unfair treatment. They argue that their foreign-owned small and artisanal mining companies are also destroying lands, forest, and water bodies in Ghana, but are in full operation with full government support.

While there are various calls for restraint, a ban, or permission to continue engaging in small-scale mining, others are advocating for a debate on the way forward on small-scale mining in Ghana. It is both the response to the call to ensure sustainable exploitation of our natural resources and the role the church can play that influence the writing this paper. The paper postulates that the church can seek for a change in human behavior—a change that begins from within—before the problem *galamsey* can be addressed.[2] The assertion is that a person's inner faith influences his or her behavior and actions.

The Biblical Concepts of Land and Stewardship

In the Bible, God is recognized as the owner of land (Gen 12:1; Exod 3:8; Lev 25:23–24; Deut 10:14; Josh 1:2; Ps 24:1–2; Jer 2:7). God created land (Gen 1:1; 2:4; Job 38:4; Ps 90:2; Isa 40:28; 42:5; Neh 9:6; Rom 1:25; Col 1:16–17; 1 Pet 4:19), cares for the land (Deut 11:12; Job 12:7–10), and has given the land to humanity to take care of (Gen 1:26; 12:7; 15:18; 17:7–8; 26:3; 28:13; Lev 25:23; Jer 2:7; 30:1–3; Ezek 11:17; Amos 9:14–15). God would destroy those who destroy the land (Gen 6:11–13; Rev 11:18). Humanity, therefore, ought to take good care of the land. Stewardship is the responsible care, use, and development of another's property. In the Bible,

2. See Bergstrom, "What the Bible Says," (Nov 14, 2014), 1.

everything in this world, including human beings, belongs to God (Gen 1:1—2:24; Ps 24:1–2). Humanity is entrusted with the care of God's property in this world (Gen 1:26–28; 2:15).

John Bergstrom emphasizes the fundamental biblical teaching that a changed human behavior arising from real conversion and from a person's inner mind, spirit, heart, and manifested in actions and attitudes, can build an ideal society.[3] As humans, the substance of our inner faith determines the living out or practice of our faith.[4] Bergstrom further explains that a Christian's environmental ethic should be based on the principles of creation value, sustained order and purpose, and universal redemption.[5] Bergstrom discusses three principles of Christian environmental ethics: first, God created the universe and therefore all his works of creation are very important to him (Principle of Creation Value);[6] second, God creates and sustains all elements and systems in his creation within particular orders to meet specific continuing purposes (Principle of Sustained Order and Purpose); and third, everything in the created world and universe is subject to corruption by sin and ultimately redemption through Jesus Christ (Principle of Universal Corruption and Redemption).[7] Thus, in agreement with Bergstrom, we add that God, who created the land, is interested in how human beings use it.

Galamsey Menace in Ghana

The term *galamsey* refers to small-scale mining in Ghana. The term is derived from a corrupt pronunciation of "gather them and sell." The name is associated with the surface mining of natural resources such as gold in Ghana. It furthermore covers illegal entry into sealed mining pits in formerly mined areas in Ghana. With abundant natural resources such as gold and diamonds in Ghana, Ghanaians as well as foreigners move in to mine these minerals in places. The industry has attracted both Ghanaians and foreigners, leading to the overexploitation of Ghanaian forests, rivers, and farmlands for gold. From 2007 to 2017 there was a surge of gold mining in Ghana.[8] There have been increases in both the numbers of workers and expansion of *galamsey* operations in Ghana.

3. Bergstrom, "What the Bible Says," 2.
4. Bergstrom, "What the Bible Says," 2.
5. Bergstrom, "What the Bible Says," 2.
6. Bergstrom, "What the Bible Says," 2.
7. Bergstrom, "What the Bible Says," 2.
8. Hilson, "Shootings and Burning Excavators," 109–16. According to Dr. Toni

These small-scale miners usually process their booty along river banks. These activities result in the pollution of water bodies, destruction of forest and agricultural lands, and loss of human life. Rivers like the Offin and Ankobra, known for massive gold deposits along their banks, have witnessed pollution with cyanide (known for gold production), resulting in a change in the color of these waters. These rivers have become unsafe for drinking. Since these rivers are sources of drinking water and other domestic use for most Ghanaians, many face potential health risks because of *galamsey* operations along these water bodies.

Another potential risk posed by *galamsey* operations has been the massive destruction of agricultural lands. Many farmers have lost their farmlands to *galamsey* operations through forceful takeovers. Some landowners sell their farmlands to these *galamsey* operators for quick cash, depriving potential and active farmers of arable lands for agriculture. Some cocoa farms and other perennial crops have been destroyed because of the activities of *galamsey* operators in Ghana. These have resulted in the loss of revenue to farmers and loss of foreign exchange to the government. It likewise deprives some farmers of food and employment.

The activities of *galamsey* operators, moreover, have destroyed forest and forestry products in Ghana. Fauna and flora in the forest are at the mercy of illegal *galamsey* operators on government lands. The destruction of forests causes deforestation, erratic rainfall patterns, drought, destruction of river sources, flooding, and destruction of houses, and also worsens the depletion of the ozone layer. The search for a permanent solution to the *galamsey* menace in Ghana, therefore, is in order. Having discussed the *galamsey* menace in Ghana, the paper proceeds to discuss sustainable development.

Sustainable Development of Land

Development refers to "expansion by process of growth."[9] Sustainable development is defined as "development that meets the needs of the present

---

Aubynn, Chief Executive Officer of The Ghana Chamber of Mines, there were about thirteen large- mining companies, sixteen operations, and over 1,000 registered small-scale companies in the sector in 2013 (See Toni Aubynn, "Mining and Sustainable Development," 3). A newspaper report indicates that there are about 3,000 registered small-scale mining companies as of 2018. Ghana National Commission for UNESCO National statistics released in 2010 indicated that the gold-mining industry was a major employer in Ghana since it employed over 520,000 Ghanaians, with about 4 percent working in the large-scale subsector and 96 percent in the small-scale mining subsector.

9. Keita, "Philosophy and Development," 132.

without compromising the ability of future generations to meet their own needs."[10] The quest for sustainable development is not an option, but a mandate.[11] Sustainable development seeks to foster economic advancement and progress without compromising the long-term worthiness of the land and its resources. It produces a platform for formulating environmental policies and its integration and development strategies.[12] The concept of sustainable development is an attempt to foster an equilibrium between the long-term stability of the environment and the economy, and to create a conducive atmosphere for economic advancement whilst upholding policies to sustain the environment.[13] Rachael Emas points out that governments and countries have realized the importance of sustainable development, resulting in policies and programs aimed at achieving clean air, potable water, and a healthy environment.[14]

Effectively, sustainable development should ensure that present development does not compromise a sustainable future of the environment. Therefore, processes of development must guarantee the preservation of resources for posterity.[15] This means that the conservation of social, natural, and manufactured capital is critical in ensuring sustainable development and integrated equity.[16]

The Sustainable Development Goals (SDGs), otherwise known as Global Goals, are "a call on the world to collectively mobilize to end poverty, protect the planet and ensure that all people enjoy peace and prosperity."[17] The SDGs came into effect in January 2016. These goals include: (1) No poverty; (2) Zero hunger; (3) Good health and well-being; (4) Quality education; (5) Gender equality; (6) Clear water and sanitation; (7) Affordable and clear energy; (8) Decent work and economic growth; (9) Industry, innovation, and infrastructure; (10) Reduced inequality; (11) Sustainable cities and communities; (12) Responsible consumption and production; (13) Climate action; (14) Life below water; (15) Life on land; (16) Peace, justice,

---

10. Emas, "Concept of Sustainable Development," 1; United Nations General Assembly, Development and International Co-operation: Environment, *Report of the World Commission*, 43.

11. See United Nations Development Programme, UNDP Support to the Implementation of Sustainable Development Goal 1: Poverty Reduction (2016), 3.

12. Emas, "Concept of Sustainable Development," 1.

13. Porter and van der Linde, "Toward a New Conception," 97–118.

14. Emas, "Concept of Sustainable Development," 2.

15. Antle and Heidebrink, "Environment and Development," 603–25.

16. Emas, "Concept of Sustainable Development," 2.

17. United Nations Development Programme. UNDP Support to the Implementation of Sustainable Development Goal 1: Poverty Reduction (2016).

and strong institutions; and (17) Partnerships for the goals.[18] According to the UNDP report, these 17 goals build on the Millennium Development Goals and also include "new areas such as climate change, economic inequality, innovation, sustainable consumption, peace, and justice, among other priorities. The goals are interconnected, often, the key to success on one involves tackling issues more commonly associated with another."[19] It goes further to state that:

> The SDGs demonstrate the combined efforts of partnership and pragmatism to make now the right choices to improve life, sustainably, for the future generations. They provide clear aims for all nations to adopt as priorities and the environmental challenges of the world at large. The SDGs are an inclusive agenda. They tackle the causes of poverty and unite us together to make a positive change for both people and planet.[20]

Thus, sustainable development ensures the protection of resources for development without sacrificing the welfare of posterity. It aims to guarantee a better future in the human quest for growth.

Implications and the Way Forward

To promote the work of the government and people of Ghana in providing sustainable development of *galamsey*, the church ought to be seen providing the light and the way. The search for a possible and lasting solution to ensure sustainable development to the *galamsey* threat in Ghana is needed and timely.

When one allows the spirit of God to rule the heart, there can be a changed heart. Without the spirit of God convicting humanity of sin, human beings are incapable of knowing they are erring. Changing one's heart is a process; it begins from conviction and it leads to salvation. By conviction, we allow ourselves to believe something different from what we formerly believed (Luke 13:3–5). Such a person exemplifies a change of mind toward old behavior in an effort to now do what is right and acceptable. With the help of God, one becomes empowered to produce both regret and change of conduct and eventually live a virtuous life.

18. United Nations Development Programme. UNDP Support to the Implementation of Sustainable Development Goal 1: Poverty Reduction. (2016).

19. United Nations Development Programme. UNDP Support to the Implementation of Sustainable Development Goal 1: Poverty Reduction (2016).

20. United Nations Development Programme. UNDP Support to the Implementation of Sustainable Development Goal 1: Poverty Reduction (2016).

The Christian church must begin a strong campaign on creation and stewardship that upholds God as the Creator and Sustainer of the universe. It must advocate that human beings are stewards of the land, and are accountable to God—both Christians and non-Christians.[21] All, therefore, ought to take good care of the land because God will hold all accountable. Proper maintenance of the environment, including the land, should be taught in our churches and places of religious meetings to ensure that members internalize this teaching and put it into practice. Effective teachings on stewardship can influence proper use of land, and make one responsible for the exploitation of land and natural resources. Understanding the Judeo-Christian tradition would enhance Ghanaians' perception and how they can take care of the land and other natural resources.

The church stands a better chance to play a mediatory role in the quest for a lasting solution to the *galamsey* menace in Ghana. In spite of Operation Vanguard, made up of the police and soldiers meant to clean up the system, illegal mining is going on. This means that the church must build on its credibility and play its role as the watchman of society and help people live with integrity and fairness. The church can influence *galamsey* operators to be mindful of the way they destroy the land and other natural resources in Ghana, and make conscientious reclamation efforts on the lands they mine for minerals. They must also be admonished to exercise restraint in exploiting and destroying water bodies and be better citizens.

The church must also encourage the government to be good stewards and to treat its citizens with respect and fairness by not transferring the destiny and property of the nation to foreigners. If the lands are put to good use by Ghanaians, there will be enough land for use so as to prevent the invasion of government and stool lands.[22] Ministers of the gospel and pastors should preach, and teach their people in authority, about the need to be responsible stewards of the land and other natural resources. Again, the church should join the government in organizing periodic workshops on the land and responsible land use. Thus, Christian churches and other religious organizations, through extensive education of the citizenry and those in authority, can help all embrace the divine ownership and human stewardship concept. It would help change people's attitudes toward land and other natural resources in Ghana.

---

21. Byrne, *Inheriting the Earth*.
22. Seth Norton, "Property Rights, the Environment," 37–54.

Conclusion

This study has explored possible solutions to curb the menace of *galamsey* operations in Ghana. The study began with introducing the extent of damage and the need for timely and lasting solutions to the *galamsey* problem in Ghana. The study linked the extent of the *galamsey* menace to the global sustainable development agenda. The study proceeded by proposing ways in which the church can play a pivotal role. It suggested that the churches in Ghana should teach their members that God is the owner of the land, and that God expects human beings to be responsible in their care, use, and development of the land. Persons in authority and the general citizenry must have a change of heart toward the environment and stewardship, and affirm that God is the owner of the land and demands responsible stewardship from humankind.

Bibliography

Antle, John M., and Gregg Heidebrink. "Environment and Development: Theory and International Evidence." *Economic Development and Cultural Change* 43 (April 1995) 603–25.

Beisner, E. Calvin, et al. "Public Policy: Environmental Stewardship in the Judeo-Christian Tradition: Jewish, Catholic, and Protestant Wisdom on the Environment-A Biblical Perspective on Environmental Stewardship." https://acton.org/public-policy/environmental-stewardship/theology-e/biblical-perspective-environmental-stewardship.

Bennett, Micah. "The Biblical Call to Environmental Stewardship." http://www.doesgodexist.org/MayJun08/BiblicalCalltoEnvironmentalStewardship.html.

Bergstrom, John C. "What the Bible Says about the Environment." https://arcapologetics.org/culture/subdue-earth-bible-says-environment/.

Byrne, Brendan. *Inheriting the Earth: A Pauline Basis for Spirituality for Our Time*. New York: Alba, 1990.

Emas, Rachel. "The Concept of Sustainable Development: Definition and Defining Principles." *GSDR* (2015) 1–14.

Hilson, Gavin. "Shootings and Burning Excavators: Some Rapid Reflections on the Government of Ghana's Handling of the Informal Galamsey Mining 'Menace.'" *Resources Policy* 54 (2017) 109–16.

Keita, Lansana. "Philosophy and Development: On the Problematic of African Development: A Diachronic Analysis." *African Development* 29.1 (2004) 131–60.

Mantey, Jones, et al. "Operational Dynamics of 'Galamsey' Within Eleven Selected Districts of the Western Region of Ghana." https://doi.org/10.22044/jme.2016.627.

Norton, Seth W. "Property Rights, the Environment, and Economic Well-Being." In *Who Owns the Environment?*, edited by Peter J. Hill and Roger E. Meiners, 37–54. Lanham, MD: Rowman and Littlefield, 1998.

Porter, Michael E., and Claas van der Linde. "Toward a New Conception of the Environment-Competitiveness Relationship." *Journal of Economic Perspectives* (1995) 97–118.

United Nations Conference on the Human Environment. *Rio Declaration on Environment and Development*. Rio de Janiero: United Nations, 1992.

United Nations Development Programme. "UNDP Support to the Implementation of Sustainable Development Goal 1: Poverty Reduction (26 Jan 2016)." https://www.undp.org/content/undp/en/home/librarypage/sustainable-development-goals/undp-support-to-the-implementation-of-the-2030-agenda.html.

United Nations General Assembly. "Report of the World Commission on Environment and Development: Our Common Future (20 March 1987)." https://sustainabledevelopment.un.org/content/documents/5987our-common-future.pdf.

Glocal Ecological Degradation of God's Gift: A Human Menace of the Divine Creation in Ghana

Yaw Attah Edu-Bekoe
Trinity Theological Seminary, Legon

Abstract

This paper looks at the great cultural mandate of creation management and stewardship in which God's creation is handed over for humanity to "till and keep it" (Gen 2:15). Throughout the Christian Scripture, from the pristine garden of Eden to the beautiful restoration in natural disclosure of John's eschatological picture of the new earth, God desires that humanity takes good care of the land. The paper therefore explores how human greed and avarice have resulted in the degradation of God's creation gift and how humanity is destroying the environment through burning, mining, and waste, while considering some biblical and theological imperatives as well as practical solutions for redeeming God's creation urges all to think globally and act locally.

Introduction

THE CENTRAL THEME OF this article is that God's created order is a gift for human heritage to keep and care for. In other words, a viewpoint of the great cultural mandate of stewardship and management of God's creation is the pivot around which this paper revolves. Three operational definitions are chosen to begin the paper. The first word is "ecology." This is the holistic explanation of the created order in the environment. In such discourse, land, for instance, does not mean the solid earth, but its totality, including the earth/soil, atmosphere, the seas, the vegetation, and water bodies such as rivers, lakes, and streams, among others. The second word is "glocal."[1] This is a cognitive term of "thinking globally and acting locally." Sadiri Joy Tira and Narry Santos stated that glocal was originally coined by Robertson "to express a new interweaving of local and global. It is simply 'thinking globally and acting locally,' as applied to business 'glocalization.'"[2] This paper considers the degradation of the created order in the global and local arenas. The third is "environment." Cunningham and Cunningham defined environment as "circumstances or conditions that surround an organism or group of organisms."[3] From the above definitions, one may proceed with the presumption that issues about ecology or environment are all in reference to one's own surroundings.[4] The environment affects human beings and people cannot avoid its impact on them, whether the impact is positive or negative. In Genesis 2:15, however, what God commands humanity to do was to keep and till the created order, otherwise known as the great cultural mandate. This means that God wanted to share the management of creation with Adam and Eve. Therefore, God has given human beings the command to oversee all of creation. The composition of the article is made up of: (a) theoretical imperatives: i.e., reflecting on the two contentions of the humanistic and evolutional foundations for the existence of the universe; (b) biblical and theological perspectives: i.e., an explanation of the mandate of management and stewardship of God's creation; (c) practical negative results of effects: i.e., challenges out of blatant disobedience to God's command of tilling and keeping his creation; and (d) praxis of meaningful solutions: i.e., opportunities for redeeming the human image from the destruction of God's creation, with concluding remarks.

---

1. Robertson, "Glocalization," 27-44.
2. Tira and Santos, "Diaspora Church Planting," 64.
3. Cunningham and Cunningham, *Environmental Science*, 14.
4. It needs to be noted that the two words are not the same. Jürgen Moltmann makes a distinction between "environment" and "ecology." See Moltmann, *Source of Life*.

Theoretical Imperatives: Humanistic/Materialistic and Evolutional Understanding of the Universe

There are two major schools of thought regarding the existence of the universe—evolutionary theory and creation theology. Since this paper is about God's gift of creation to humanity, and how humanity has from greed and selfishness destroyed and degraded this gift, a return to the theoretical imperatives is significant. So this paper has taken a preferential option for creation theology over evolutionary theory. An overview of the humanistic/materialistic, as well as evolutionary theory contentions, are presented.

Humanistic/Materialist Contention

There are many worldviews concerning the ecology/environment. Peoples' negative or positive attitudes to the ecology can be seen based on how they treat it. John Stott argued on the views people hold about ecology, including the humanistic or materialistic view. The materialistic opinion denies a Creator of the universe and the distinctive spiritual aspect of human beings, but affirms an unbounded optimism in humanss ability to solve their own problems. This can be done through using technology wisely to control the environment, conquer poverty, reduce diseases, and extend our lifespan. Thus, for those who hold the materialistic view, the environment is there to be exploited for human benefit. But Stott sounds a caution that exhausting our natural resources will bring about destruction to humanity.[5]

The humanistic view is that nature is simply there. This means the existence of nature and all its resources is taken for granted. This has the implication that if energy is uncreated, then there is no Creator. Therefore, no divine imperative is to be used in a particular way to check people's attitudes toward the environment."[6] Indeed, Norman Geisler was right when he criticized humanists thusly:

> It is scientifically inappropriate to claim that energy cannot be created or destroyed. The first law talks of the contestants of the universe but does not make any statement about its origin. Moreover, the second law of thermodynamics speaks about the origin of the universe because it is able to determine that [the] amount of usable energy in the universe is decreasing. It is winding down as a result of it being wound up. This suggests that the universe is not eternal and if it is so, then it must have

---

5. Stott, *Issues Facing Christians Today*, 135.
6. Brown, "Issues Facing Christians Today," 135.

been created by a Creator . . . The bible affirms that "In the beginning God created the heavens and the earth" (Genesis 1:1).[7]

Another assertion of the humanistic view is that technology can solve all our problems if it is used wisely, and even those problems that technology cannot solve can be solved by governments. Geisler refuted and contended that technology cannot solve all problems since we cannot know all the relevant facts to solve our problems. Even if people know all the relevant facts, they still cannot avoid errors. He stated that we deceive ourselves when we think that we are omnipotent and can do everything. Our knowledge and control of the future is limited, and our inventions to do things right hardly work or get the desired results, and "if they work at all, in ways that we do not expect, our planning is meaningless, our systems are running amok—in short, the humanistic assumptions upon which our societies are grounded lack validity."[8] Indeed the materialistoc view about the environment cannot be trusted because the more advanced technology is put in place and the more sophisticated and numerous government institutions are set up by humanity, the more human beings fail to deal with the problems of the environment.

Evolutionary Theory Contention

The pioneer of evolutionary theory was Charles Darwin, who indicated that human beings are an inherited heritage from the species of apes. This means in simple terms that human beings are a higher development of the monkey species. In his discussion on "The Origin of Man," Louis Berkhof made the following statement to explain the theory of evolution:

> According to thorough-going naturalistic evolution, man descended from the lower animals, body and soul, by a perfectly natural process, controlled entirely by inherent forces. One of the leading principles of the theory is that of strict continuity between the animal world and man. Nothing that is absolutely new and unpredictable can appear in the process. What is now found in man must have been potentially present in the original germ out of which things developed, and the whole process must be controlled from start to finish by inherent forces. Theistic evolution . . . simply regards evolution as God's method of working . . . It sometimes held that only the body of man is

---

7. Geisler, *Christian Ethics*, 315.
8. Geisler, *Christian Ethics*, 315.

derived by a process of evolution from lower animals, and that God endowed this body with rational soul.[9]

This viewpoint has been cited for a couple of reasons. First, this article is skewed toward Christian creation theology; thus, no extensive development in the theory of evolution can fit the scheme. Second, even evolutionary theorists have among them those who are theistic evolutionists, even though part of its scholarship is accepted by some theologians. This goes ahead to affirm the fact that the hypothesis for evolution is not sacrosanct; it is violable and has not been conclusively accepted worldwide. Berkhof posits that the intelligent design movement "argues that the biosphere is possessed of an irreducible complexity: which makes it impossible to explain its origins and development in any other method than positing intelligent design."[10] Bosch, in discussing mission in the wake of the Enlightenment, recorded something about the humanist:

> By 1917, Walter Rauchenbusch, a major exponent of the Social Gospel, could confidently declare that the doctrine of the kingdom of God was "itself the social gospel" . . . This meant, in effect, the discarding of all supernatural features. Reality was entirely inner-worldly, anthropocentric, and naturalistic. "Is anything in the whole universe of God, when rightly understood, supernatural?" asked W. B. Brown . . . The miraculous was eliminated and superseded by professionalism, efficiency, and scientific planning.[11]

Hence, it is by faith that God created the earth and humanity cannot understand God in all things. This indicates that the humanistic/naturalistic/materialistic and evolutionary theory are modern and postmodern ideas from the Enlightenment. In fact, it is in support of God's intelligent design that this article is skewed toward the preferential option for creation theology.

Biblical and Theological Perspectives

Roger Greenway posited that the great cultural evolution is a composite theology of celebrating God's glory in creation. He submitted that the great cultural mandate of stewardship is in several parts.[12] Greenway posits these

9. Berkhof, *Systematic Theology*, 183, 184.
10. Berkhof, *Systematic Theology*, 184.
11. Bosch, *Transforming Mission*, 321.
12. In the reflection and celebration of the mandate: The last part of the cultural

three mandates. The first was to be fruitful. God commanded man in this cultural mandate to be fruitful, increase, and fill the earth (Gen 1:28). This was the basic command of building community, and this God commands humanity to do through marriage and family (Gen 2:24). This indeed was the cultural foundation for society. The second was the naming of creation (Gen 2:19). This mandate given to Adam was to name all that God has created. Here, the human being's mental and aesthetic abilities were tested. As Adam named all of God's creation, he was responsible to unlock and see the secret potential in things. Naming all manner of things and animals revealed the beauty and variety in God's creation. And the final mandate was tilling and keeping creation (Gen 2:15). After creating man, God placed him in the beautiful garden of Eden to till and keep. This was to supply man with all his physical needs as he tilled it so that his descendants would also benefit from it. The human being was therefore asked to manage and maximize God's creation or his personal benefit. Here is the relevance of the inclusion of the "keeping it" aspect of the mandate. Keeping God's creation means humanity must not abuse, exploit, or misuse it. In all, Scripture is the main source to turn to in order to expatiate the creation theology.

Biblical Doctrine and Theology of God Creating Ex Nihilo

Wayne Grudem also defined the doctrine of creation thusly: "God created the entire universe out of nothing; it was originally very good; and he created to glorify himself."[13] Grudem went on to say, "When we say that the universe was created 'out of nothing,' it is important to guard against a possible misunderstanding. The word nothing does not imply some kind of existence, as some philosophers have taken it to mean. We mean rather that God did not use any previously existing materials when he created the universe."[14] Grudem further indicated five areas relevant for the discourse:

- God created the universe out of nothing
- Creation is distinct from God yet always dependent on God

---

mandate includes the element of reflection and celebration of God's creation. After God had created the earth, firmament, seas, animals, and the like, God saw all that he had done and declared that his creation was "very good" (Gen 1:31). God took a day to rest, to reflect, and to enjoy his own creation. Thus, he commands us to honor the Sabbath, that is to say, to take a day of rest, reflect, and celebrate God's creation in worship. See Greenway, "Cultural Mandate," 251.

13. Grudem, *Systematic Theology*, 262.
14. Grudem, *Systematic Theology*, 263.

- God created the universe to show his glory
- The universe God created was "very good"
- The relationship between Scripture and the findings of modern science.[15]

On creation out of nothing, Grudem stated:

> The Bible clearly requires to believe that God created the universe out of nothing. (Sometimes the Latin phrase *ex nihilo*, 'out of nothing' is used; it is then said that the Bible teaches creation *ex nihilo*). This means that before God began to create the universe, nothing else existed except God himself. This is the implication of Genesis 1:1, which says, "In the beginning God created the heavens and the earth." The phrase "the heavens and the earth" includes the entire universe. Psalm 33 also tells us, "By the word of the Lord the heavens were made, and all their hosts by the breath of his mouth . . . For he spoke, and it came to be; he commanded, and it stood forth" (Psalm 33: 6, 9).[16]

What can African theology derive from such biblical and theological essentials? As J. O. Y. Mante posited, there are two main reasons why the whole ecosphere is to be taken seriously in any African theological consideration:

1. The nonhuman environment occupies a major portion of both the life and thought of the majority of Africans. The nonhuman environment pervades African symbolic thoughts. The neglect of the nonhuman environment may be due to foreign cultural influence.

2. Even if it were the case that the nonhuman environment did not play a major role in the lives of most Africans, it is common knowledge that the current destruction of the ecosystem is leading to starvation and less quality of life, and must therefore attract our theological discussions.[17]

These views imply that there are many ecological challenges that the world is now facing as a result of poor management of our environment. It is well known that such challenges have brought about global warming, pollution, diseases, and flooding, to mention but a few. For Schonberg, ecology is about relationships. He posited:

---

15. Grudem, *Systematic Theology*, 262–73.
16. Grudem, *Systematic Theology*, 262, 263.
17. Mante, *Africa*, 4, 5.

> Ecology does not begin with pop cans and pesticides but values, goals, the meaning of life, and why we were on that road in the first place. Ecology deals with relationships—the totality of relationships between living organisms and their environment. Life in general is no more nor less than the sum of its relationships. How we approach these relationships depends on what we really believe. The kind of relationships we get speaks to the validity of our beliefs. Beautiful relationships tend to confirm. Strained and broken relationships tell us something else.[18]

Schonberg is right in making this assertion that a good relationship with the environment must lead to a good relationship with God. And to be more responsible, one must think globally and act to help the local.

Practical Negative Realities of Effects

Glocal Depletion of the Ozone Layer by Both MDCs & LDCs

A typical *glocal* instance is the destruction of the ozone layer. What is this ozone layer? In my view it is God's created gift of a cover in the atmosphere to prevent ultra-heating or infra-red ray transmissions from the sun. Gaseous emissions from human activities have created dangerous holes in the ozone layer, thus creating the currently unsolvable condition of global warming on earth. Snow cover which has remained on the land for some time in some parts of the world is melting, especially in both the Arctic and Antarctic regions. The Less Developed Countries (LDCs) are the worst affected. The Sahara Desert and the Sahel region grasslands are believed to be spreading into the sub-Saharan African forest areas at an alarming rate. In addition, the weather in Africa as a whole has changed, with erratic rainfall patterns leading to drought, famine, and death on a large scale.

On the global level, the More Developed Countries (MDCs) have caused lots of global casualties through their developmental processes since the Industrial Revolution. These MDCs are mainly in Europe, North America, Southeast Asia, Australia, and New Zealand. Over the years, these MDCs, due to developmental processes, have built manufacturing industries and other factories that have aided the gradual depletion of the ozone layer. Industrial productions such as textiles, locomotives and other railway machines, automobiles, jets, space exploration tubes and launches, nuclear

---

18. Schonberg, *Ecology and Beyond*, 7.

tests and bombings, and other chemical warfare bombs, among others, have gradually and persistently depleted and created holes in the ozone layer.

Local Depletion of the Ozone Layer and Dangerous Landfill Burning

On the local level, apart from the manufacturing industries that emit gasses into the atmosphere, the gas fumes from dilapidated vehicles, the burning of plastics and vehicle tires on dumpsites which can be likened to the biblical *gehena*, as well as the tree burning for charcoal, among other things, are helping to deplete the ozone layer.

The use of plastics has increased over the years due to affordability, durability, and flexibility. Plastics have become a menace, and to dispose of them we tend to burn them at dumpsites. Burning plastics is dangerous because it has been proven that burning plastics is not without harmful consequences to human health, and messages about the menace of plastics have gradually but persistently taken the headlines. Messages going around on social media suggest the use of plastics in cars and food contact materials, among other things, "are a contributing factor to the increase in cancer cases around the world."[19] It is understood that when such plastics are exposed to heat such as is associated with hot food and prolonged exposure to the sun, benzene is emitted into the environment. The message concludes that the benzene released by this process endangers the health of users and causes cancer. Some scientists claim that this is a misconception drawn from misrepresented facts. Actually, Gensler has stated: "The anonymous originator of this message demonstrates little knowledge of chemistry."[20] The American Chemistry Council also posited:

> The message is seriously flawed when it suggests that plastics are the means by which benzene is introduced into the atmosphere. Benzene occurs naturally in the environment. In addition, human activities like tobacco smoking, automobile fumes, and gasoline transfer contribute heavily to the benzene in the atmosphere. There is benzene all around us, with or without plastics. The seriousness of poisoning caused by benzene depends on the amount, route, and the length of time of exposure, as well as the age and pre-existing medical condition of the exposed person.[21]

---

19. Mikkelson, "Do You Know the Danger?," para. 13.
20. Gansler, "Is Your Car Killing You?," para. 4.
21. American Chemistry Council, "Benzene," para. 12.

The above reinforces the view that burning is dangerous and harmful. In addition, as Castle et al. indicated, though some chemicals can cause cancer, it also known that benzene levels in food contact materials such as plastic spoons, cups, and plates made from polystyrene, are very low.[22]

The location of the nonfunctional dumpsite at Pantang in the Greater Accra Region, close to a busy road and a hospital, is still not a good scene. Moreover, there is a perpetual fire emitting smoke and gasses into the atmosphere there. In addition, the stench from the site, as well as the smoke, has been disturbing hospital patients, residents, drivers, and passersby. The unfortunate unhealthy nature of this site is twofold:iIt is close to a lot of people, and it is very close to the Pantang Mental Hospital.

When it comes to the use of plastics then, benzene exposure may not be our main problem. The problem is with harmful chemicals that are released when such plastics are subjected to incineration or burning, such as what happens at the above landfill. Most people who burn their domestic plastic waste do not realize how harmful this practice is to their health and to the environment. A study indicates that "backyard-burning of waste is far more dangerous to our health than previously thought."[23] In Ghana, this risk is compounded with the daily use of plastics for lighting fires (for cooking) and landfill burning across the country. "It can increase risk of heart disease, aggravate respiratory ailments such as asthma and emphysema, and cause rashes, nausea, or headaches, damages in the nervous system, kidney or liver, or in the reproductive and development system."[24] The burning of polystyrene polymers such as foam cups, meat trays, egg containers, and "take away" packs, among others, releases styrene. "Styrene gas can readily damage the skin and lungs. At high levels, styrene vapor can damage the eyes and mucous membranes. Long-term exposure to styrene can affect the central nervous system, causing headaches, fatigue, weakness, and depression. This exposure extends through the environment to affect neighbors, children, and families."[25]

Moreover, the most dangerous emissions can be caused by burning organochlorine-based substances like PVC. As WHO stated:

> When such plastics are burned, harmful quantities of dioxins, a group of highly toxic chemicals are emitted. Dioxins are the most toxic to the human organism. They are carcinogenic, a hormone disruptor and persistent; they accumulate in our

---

22. Castle et al., "Migration of Plasticizer Acetyltributyl Citrate," 914–20.
23. Graham, "EPB 433," para. 3.
24. Graham, "EPB 433," para. 2.
25. Graham, "EPB 433," para. 11.

body-fat and thus mothers give it directly to their babies via the placenta. Dioxins also settle on crops and in our waterways where they eventually wind up in our food, accumulate in our bodies, and are passed on to our children. Thus, the effect of burning such plastics is not limited to the 'offending' generation but even to generations unborn.[26]

Burning, however, as a means of disposing of plastics, is a grave health and environmental hazard. Unfortunately, owing to ignorance and the absence of proper disposal options in LDCs of the world such as Ghana, burning is predominantly the preferred means of disposing plastics.

Hazards of Dredging or Surface Mining in Ghana: Galamsey

Ghana, like most countries, has been a mining country for a long time. Between the time when Portuguese expeditions led by Don Diego d'Azambuja arrived at Shama (1471) and Elmina (1482), and 1957, when Ghana had its independence, the nation was known as the Gold Coast because of the large deposits of gold found in the land. Boateng mentioned three main ways in which gold mining operations are carried out in Ghana—underground mining, open-cast mining, and dredging:

> Underground mining, the most common method of the three is for two types of gold. These are the "blanket conglomerate"' and "lodes." With this method, the miners go down into great depth as much as 3,000 feet underground, use explosives to break up the ores and send them to the surface for crushing and extracting of gold. Open-cast mining is much simpler: it consists of digging up the ore by mechanical excavators and sending it to mills for crushing and extraction of the mineral. For dredging, which is mainly done in rivers, large dredges bring up sand rocks from the bottom of the river, wash them, and separate little grains of gold contained in them.[27]

Significantly, Ghanaians and foreigners together engage in mining in Ghana. An interesting report on the operations of foreign companies engaged in underground mining was published in 2018 in the Ghanaian Times by Yaw Kyei titled, "President Slams Foreign Mining Firms . . . for Deplorable State of Mining Communities." He wrote:

26. World Health Organization, "Dioxins and Their Effects," para. 6.
27. Boateng, *Geography of Ghana*, 94.

> The President . . . has urged multi-national mining firms in the country to be actively involved in the development of the communities where they operate. The President said the firms had been mining in communities such as Obuasi, Tarkwa, Prestea, among others, for decades but the state of mining communities were very deplorable. "Why is Obuasi not the most beautiful Community in Ghana? . . . Why does Akwatia's appearance not reflect anything about the diamonds that have been taken from the soil all these years? The distressed state of mining companies is nothing short of disgrace and we must work to change the situation . . ." He urged the companies to commit to the development of the communities where they were making their monies and help improve the living conditions there.[28]

What this means is that the wealth from the minerals is not translating into the development of the communities. Of late the menace that has devastated the nation is surface mining and/or dredging, otherwise known by Ghanaians as *galamsey*. What is *galamsey*? The etymological meaning of *galamsey* is "gather them and sell." This means digging to collect in bits through small-scale mining to search for wealth, such as gold and diamonds. The methods used in mining are harmful to the environment because the top soil of the earth's surface is massively removed and the bottom of river bodies are dredged before the gold ore is extracted. By so doing, it affects the vegetation, animal life, and aquatic life, among other things. Francis Osei Owusu recalled the history of *galamsey* at Pokukrom:

> In 1992, the operations of Dunkwa Goldfield Limited collapsed and this paved the way for small scale mining operations popularly known as *galamsey* at Pokukrom. It was mainly done by indigenous people of the town. In this operation sand was obtained from the river bed, water ways and streams. These were then washed and the tiny particles of gold in it were gathered. As the etymology of the name *galamsey* goes "gather them and sell," the people wash the sand, collect the tiny particles of gold in the sand together and sell them to the gold dealers.[29]

It may be observed that as sophisticated machines were introduced, the number of people involved has increased, and the rivers, streams, and the land have further become destroyed by deep excavations. The greedy, covetous, and negatives attitudes of foreigners such as the Chinese have created havoc and massive degradation of the environment, leading to the poor

---

28. Kyei, "President Slams Foreign Firms," 16.
29. Owusu, "Environmental Degradation an Issue," 31.

socioeconomic lives of the indigenous Ghanaians. Nevertheless, these illegal operations of these foreigners could never have succeeded without the massive support of indigenous politicians, traditional leaders such as chiefs and kings, local opinion leaders, and the common masses who struggle to get their daily bread.

Negative Attitudes toward Trash Resulting in Persistent Flooding

On June 3, 2015, it rained in the night and many areas in Accra became flooded, resulting in a gas explosion and the burning of the Nkrumah Circle Goil fuel filling station and its surroundings. The fire-water flooding and disaster killed about 200 Ghanaians. Almost three years later, on Thursday, May 17, 2018, a rainstorm pummeled the Akuapem ridge and its low-lying Accra pains. The clouds indicated that the storm originated from the sea and it lasted for about two hours (11:00 am to 1:00 pm). This rain also caused another flood and some deaths were recorded. The causes for these floods are commonly attributed to the menace of filth, trash, and plastics, among other things, all of which create a dirty environment, and choked trenches and gutters, as well as persistent flooding. There is also the spread of diseases such as cholera, diarrhea, and malaria.

The overall effects of these three examples of *glocal* devastation of the environment in Ghana are very enormous and diverse. A representative list of these devastating effects are:

1. Pollution of air, water, and noise, which include oxides of nitrogen, sulphur carbon, and dust in the atmosphere;

2. Degradation of the land through filth, which destroys farmlands, cash, and food crops, and leads to loss of life;

3. Negative attitudes, which destroy the environment through the resultant filth; and

4. Despoliation, deforestation, and destruction of land surfaces through the activities of the *galamsey* operators.

With regard to the third and fourth points, there is the need to elaborate with some local reports. On May 22, 2018, Vision One News gave Nana Osei Marfo's report which stated that the "Densu River is dead."[30] The report talked about trash, buildings on waterways, the encroachment of rivers on

---

30. Vision One News, "Densu River is Dead."

the banks, throwing human wasts into the river, and *galamsey* discoloring waterbodies through the discharge from mud. It also mentioned acid pollution, effluents, and toxic substances such as selenium, fluorine, nickel, and radi-active elements. Instances of humans bathing with pigs in the Densu River, flowing into the Weija Dam for treatment for human consumption (i.e., the facility that supplies potable water to the people living in the West of Accra), were also cited. This report indicated that Oduro Ampaw, an assembly member of the New Weija Electoral Area, made a statement for the assembly to do something about the Densu River dying. The contractor who was working on the Weija Dam was charged to supply dustbins to individual houses to prevent people from throwing trash into the river. In addition, thousands of oil palm seedlings were to be supplied and planted along the banks of the river. At present, little has been done.

Again, on May 25, 2018, Opera News reported, in an article entitled, "*Galamsey* Queen's Partner Missing," that:

> State attorneys are looking for a Chinese Lu Qin Jun who is an accomplice of Aisha Huang, the Chinese lady (tagged *galamsey* queen) who is standing trial over alleged illegal mining in Kumasi . . . Aisha Huang and four other Chinese nationals—Gao Jin Cheng, Lu Qi Jun, Habin Gao, and Zhang Pen—filed the application two weeks ago praying the court to show clemency and review some of the bail conditions which they claim were posing financial challenges to them. The five were arrested in Kumasi for their alleged involvement in large-scale illegal mining and were brought to Accra to face trial. Aisha was granted bail by the court in June 2017 in the sum of GHC 500,000.00 with two sureties to be justified. the four others were also granted bail in the sum of GHC 500,000.00 with surety each who must be Ghanaian by birth.[31]

Nonetheless, some other Chinese institutions are prepared to help Ghana solve the challenge of *galamsey*.[32] This means that while some Chinese are involved in the destruction of waterbodies and degradation of the environment through *galamsey* surface mining, there are official Chinese actions being taken to help Ghana resolve its land degradation challenges.

---

31. Opera News, "Galamsey Queen's Partner Missing," para. 11.
32. Kale-Dery, "China Willing to Support Ghana," 23.

Praxis of Meaningful Solutions for Redeeming God's Creation

There are many ecological challenges that the world is now facing as a result of poor management of our environment. It is well known that such problems have brought about pollution, diseases, floods, and global warming, just to mention but a few problems. How can the world reverse the current environmental disaster? What practical opportunities can Ghana use to redeem God's gift of the Created Order that has been destroyed over the years?

Praxis for the Emancipation of God's Creation

I seek to suggest six actions for redeeming the human image from the destruction of God's creation through the massive blatant disobedience of tilling and keeping the ecosystem in Ghana.

1. *Glocal Application of Recycling: Scientific and Technological Utilization of Trash*

As defined from the outset of this article, *glocal* simply means thinking globally and acting locally. Furthermore, Goldfarb, in discussing the actual recycling question, submitted that:

> The lack of long-range planning coupled with skyrocketing disposal costs created a crisis situation in municipal waste management in the 1980s . . . Recycling, which until recently has been dismissed as a minor waste disposal alternative, is being encouraged as a major option. The Environmental Protection Agency (EPA) and several states have established hierarchies of waste disposal technologies with the goal of using waste reduction and recycling for as much as 50 percent of the material in the waste stream. Several environmental groups are urging even greater reliance of recycling, citing studies that show that more than 90% of municipal waste can theoretically be put to productive use if large-scale composting is included as a component of recycling.[33]

I seek to suggest in this article that much attention should be given to Goldfarb in an LDC like Ghana. Separate receptacles for bottles, cans, cardboard, and papers, among others, should be provided by trash companies such as Zoomlion. These would then be collected by trash companies for

---

33. Goldfarb, *Taking Sides*, para. 9.

recycling and for various uses. This means that recycling companies must be encouraged in the country. Ghana can use this scientific and technological way of dealing with its trash. For instance, in a news item titled, "KMA Inspects Compost Facility in Kumasi,"[34] it was reported that the Kumasi Metropolitan Assembly (KMA) has begun operating its compost facility in Kumasi to use waste to produce fertilizer for agriculture, and paper waste to produce toilet rolls. Recycling will go a long way to solve the plastic and trash challenges in Ghana.

2. *Endorsement, Ratification, and Enforcement of International Communiqués*

International bodies such as the UN, EU, and the AU have raised concerns about the issues and have come out with various possible solutions to deal with the ecological challenges, but to no avail. Some conferences held to deal with environmental challenges include the UN Conference on Environment and Development (UNCED) held in Rio de Janeiro, Brazil in 1992, the EU Summit held in Brussels in 2009,[35] and the AU Summit held in Ethiopia in 2011.[36] As stated by Fritsch,

> Perhaps the unique contribution of the prophetic voice today is in continuity between ecojustice (justice for the whole order of created things and beings) and social justice. It cautions that a national energy plan must be one where none make an unfair sacrifice and none reap unfair benefit; but it adds that this is impossible while inequities exist among those who control the supply and information. Vested interests make any fairness principle a sham.[37]

The preservation of the environment requires attitudinal change rather than just enacting decrees and inventing technologies to deal with environmental degradation. It is important for the church to use its mission to humanity as a means to teach people about the importance of preserving the environment. The fact of the matter is that rules, conventions, and resolutions concerning the environmental preservation would be ineffective unless they are applied by human beings with a human face and godly approach. Unfortunately, decisions, resolutions, and ratifications were made during these conferences which individual nations and governments now have to implement and enforce.

34. UTV News Item, "KMA Inspects compost Facility in Kumasi."
35. Trayner, "Climate Change, Brussels."
36. CablegateSearch, "Ethiopia and Climate Change."
37. Fritsch, "Theological Foundations for an Environmental Ethics," 299, 300.

3. Institution of Deep Ecology: "Green Republic or Environment Project"

One of the major initiatives to curb ecological degradation is the adoption of a deep ecology project. As Devall cited from Brower and Glasser,

> Deep ecology perspectives and the DEM have contributed to the development of ecophilosophy, echopsychology, and intellectual discussions in particular by helping people articulate and develop their own ecosophy both individually and as part of a community . . . There are those who see hope for the future of *Homo sapiens* living with the rest of nature. They maintain that *Homo sapiens* have the capacity to develop into mature human beings both as individuals and collectively if humanity practices CPE on the earth—conservation, preservation, restoration. However, how the planet as an interdependent ecosystem, subject to increasing and generally negative human interventions, will fare in the 21st century remains an open question.[38]

This idea of deep ecology, as well as the sanitation drive, has caught the eyes of officials in Ghana. As Yakubu Aboul Majeed reports from Tamale that the Green Republic Project, aimed at planting twenty million trees in the next ten years across the country to help fight the adverse effects of climate change, was launched to help safeguard a healthy future environment. It was said that they were concerned about the rate at which Ghana's forests were being depleted and the dire consequences that deforestation poses for future generations.[39] In a related development, Ken Osei Mintah also reported that Norway has pledged support to Ghana's sanitation drive. He stated:

> The Norwegian Ambassador to Ghana has pledged his country's commitment to support Ghana to achieve the goal of making Accra the cleanest city in Africa by 2020. The support, he said, would be a technical cooperation between the Ministry of Environment, the Environmental Protection Agency (EPA) and their corresponding ministries and agencies in Norway . . . This was when Young Reporters for the Environment (YRE), a body aimed at empowering young people to take a stand on environmental issues, and Green Youth Organization partnered the Norwegian Embassy to engage in a beach clean-up exercise . . .[40]

---

38. Devall, " Deep, Long-Range Ecology Movement," 35. See also Brower, *Let the Mountains Talk* and Glasser, "Naess's Deep Ecology Approach," 157–87.

39. Majeed, "Green Republic Project," 3.

40. Mintah, " Norwegian Prime Minister in Ghana," 16.

4. *Utilization of Environmental Justice: Strict Enforcement of Leadership Normative*

Most of the time, especially in Sub-Saharan Africa, research, decisions, and normative reports abound on environmental degradation and redemption, but the implementation of such reports are the challenge. For example, Bill Devall wrote about the Global 2000 report in the USA that much of the scientific research advanced during the 1970s, which was proclaimed the "Decade of the Environment" by President Richard Nixon.[41] He also summarized a report authorized by President Jimmy Carter and published in 1980, "Global 2000 Report to the President: Entering Twenty-First Century (CEQ 1980)."[42] This report concluded that if present trends continue, more cities will be crowded, more places polluted, and hundreds of millions of the desperately poor will be deprived of food and other necessities.[43] Devall concluded, "The continuing collective efforts to change human behavior to forestall global warming indicates that some attempts at effective political action in the face of a 'global environmental crisis' are being made."[44]

Various governments in sub-Saharan Africa have taken strong stances to deal with environmental degradation. However, making such decisions in Africa has not been easy. Yaw Kyei reported that the President of Ghana has urged multinational mining firms in the country to be actively involved in the development of the communities where they operate and that the government was in the process of streamlining the activities of the small-scale miners to mitigate the adverse effects of mining of the environment.[45] In another article, Severious Kale-Dery, back from China, stated, "The Guangzhou Province in China is willing to use its experience in addressing surface mining to help Ghana turn the '*galamsey*' communities into agricultural lands. The Guangzhou Province has high expertise in surface mining and

---

41. Devall, "Deep, Long-Range Ecology Movement," 20.

42. Devall, "Deep, Long-Range Ecology Movement," 20.

43. As Devall indicated, "The Global 2000 report was intended as a warning to humanity to collectively change its behavior . . . A convergence of various trends has led to what is frequently called the "environmental crisis." On a finite planet for expansion . . . Yet population has continued to grow, per capita consumption has increased . . . While some people believed that humans will find solutions to many problems through technology, the pace of technological changes continues to disrupt the lives of hundreds of millions of people . . . The process of worldwide economic integration, called globalization, continues to disrupt the social and economic security of billions of people while global warming, acid rain, destruction of the Ozone layer, and other industrial civilization undermine the integrity of natural systems across the planet." See Devall, "Deep, Long-Range Ecology Movement," 20, 32.

44. Depledge, "Coming of Age at Buenos Aires," 15.

45. Kyei, "President Slams Foreign Mining Firms," 16.

once suffered land degradation as a result of *galamsey* activities."[46] Although these official activities are strongly taken, still the current government of Ghana is having a tough time enforcing such normative sanctions against those illegally involved in the menace of surface mining, even after the ban and reforms in 2017/2018.

5. *Pedagogical Mentoring: Catching the Positive Attitudinal Change Young*

The adage of "catching them young" is profound for Ghana. This means the young can be good role models in environmental redemption if they are educated about it earlier. As such, teaching children and youth in civic education in both public and private schools as well as in their homes and in the church is something to pursue. After all, "cleanliness is next to godliness." As reported by Kukua Nketsia Asiedu,

> The Ga East Municipal Assembly has launched an initiative known as the Green View Environmental Club, to engage school children to fight against environmental abuse in the municipality. Led by the Municipal Chief Executive (MCE) . . . the club will educate schoolchildren on the proper ways to handle waste . . . She called on stakeholders, including traditional rulers, pastors, imams, and teachers, to join hands in the effort to rescue the country from its present situation where it was overwhelmed with filth.[47]

I seek to suggest that the civic education in our schools in the early and mid-1960s and 1970s should be reintroduced. There was also Abbor Days when school children planted trees. This should be reimplemented. The days when high schools and colleges planted their own food crops should be reintroduced. Yet, another commendable news item indicated that Ghana has launched a project dubbed "Green Ghana"[48] for the next ten years. The project will encourage tree planting, biodiversification, and the control of desertification. This was in the launching of forestation in Ashanti Region by the Forestry Authority on August 12, 2018. The idea is to disseminate such information and programs down to the young ones and into the grassroots as a national process of stemming the downward trend of desertification.

---

46. Kale-Dery, "China Willing to Support Ghana," 15.
47. Asiedu, "Ga East Assembly Launches Green Environment Club," 16.
48. Adom TV News Item, "Green Ghana for the Next Ten Years."

6. Holistic Biblical Anthropocentrism: Obedience to God's Creation Management Command

In his paper on dialectical anthropology, Simkins asked the question, "Can religion contribute to better, more ecologically balanced treatment of the environment?"[49] Simkins's paper challenged the common anthropocentric reading of the Bible. He then contended instead,

> That the Bible is a product of a theocentric worldview. Humans may be singled out in the Bible for particular attention, but they are not separated out from the natural world in which they live. In the theocentric worldview of the Bible humans and all other creations are dependent upon God for creation and subsistence, and all alike are valuable to God as part of his creation. The world inclusive of humans and animals, trees and plants, land and seas, belong to God because it is God's creation.[50]

Simkins's conclusion is relevant for this article because it expanded on the theocentric concept and biblical anthropocentricism, and proposed an attitude of human beings having more in common with other living creatures such that we are obedient to God's creation stewardship and management command. He posited, "The world belongs not to humans but to God, who created and sustains it . . . In a theocentric worldview, humans have more in common with other living creatures than they have differences. All alike are dependent upon God for creation and subsistence and all alike are valuable to God as part of his creation."[51]

It is quite unfortunate that the human being, the apex of God's creation, out of greed and selfishness, has blatantly refused this command of stewardship, which has led to the global ecological degradation of God's creation. Ghanaians are no exception.

Conclusion

This paper first started with theoretical imperatives of two contentions for the humanistic and evolutionary foundations for the existence of the universe. Second, the biblical and theological explanations of the mandate for management and stewardship of God's creation were explained. The third part included the challenges that have arisen out of blatant disobedience to God's command of tilling and keeping his creation. Finally, the article

---

49. Simkins, "Bible and Anthropocentrism," 397–413.
50. Simkins "Bible and Anthropocentrism," 397.
51. Simkins "Bible and Anthropocentrism," 412.

considered the opportunities for redeeming humans' image from the destruction of God's creation.

Significantly, the fourth great commandment of witness, that is the Markan narrative, states: "Go into all the world and preach the good news to all creation (Mark 16:15). A simplified exposition of this text is that while Christians are doing their part to rescue the perishing, they should be able to preach the good news about creation care to salvage or redeem the degraded ecology. In other words, "to keep and to till it" is an ecological mandate for stewardship and management which must be preached by the churches. Believers dare not fail Christ, our Lord.

Bibliography

Adom TV News. "Green Ghana for the Next Ten Years." www.igreengrowthsolutions.com/.../news/30-iggs-kas-push-climate-change-adaptation.

American Chemistry Council. "Benzene." http://chemicalsaftyfacts.org/benzene.

Asiedu, Kukua Nketsia. "Ga East Assembly Launches Green Environment Club." *Daily Graphic*, June 1, 2018. https://www.graphic.com.gh/news/general-news.html?is_preview=on&template=sportsnew&start=4880.

Berkhof, Luis. *Systematic Theology*. East Peoria, IL: Versa, 2005.

Boateng Ernest. A. *Geography of Ghana*. Cambridge: Cambridge University Press, 1996.

Bosch, David J. *Transforming Mission: Paradigm Shifts in Theology of Mission*. Maryknoll, NY: Orbis, 2005.

Brower, David. *Let the Mountains Talk, Let the Rivers Run*. San Francisco: Harper Collins, 1995.

Brown, Mark Malloch. "Issues Facing Christians Today." In *Issues Facing Christians Today*, edited by John R. W. Stott, 18–32. Grand Rapids: Zondervan, 2006.

Castle Laurence, et al. "Migration of Plasticizer Acetyltributyl Citrate from Plastic Film into Foods During Microwave Cooking and Other Domestic Use." *Journal of Food Protection* 51 (2009) 914–20.

Cunningham, William P., and Mary Ann Cunningham. *Environmental Science*. New York: McGraw-Hill, 2010.

Denison, Richard A., and John Ruston. "Recycling Is Not Garbage." *Technology Review*. October 1997.

Depledge, Joana. "Coming of Age at Buenos Aires: The Climate Change Regime in Kyoto." *Environment* 41.7 (Sept. 1999) 15–27.

Devall, Bill. "The Deep, Long-Range Ecology Movement 1960–2000: A Review." *Ethics and Environment* 6.1 (Spring 2001) 18–41.

Edu-Bekoe, Yaw Attah, and Enoch Wan. *Scattered African Keep Coming: A Case Study of Diaspora Missiology on Ghanaian Diaspora and Congregations in the USA*. Portland, OR: Institute of Diaspora Studies, 2013.

"Ethiopia and Climate Change." http://www.Cablesearch.net/cable.phd?id.

Fritsch, A. J. "Theological Foundations for an Environmental Ethics." In *Ethics in Forestry*, edited by Lloyd C. Irland, 296–312. Portland, OR: Timber, 1994.

Gansler Ted. "Is Your Car Killing You with Benzene?" http://blogs.cancer.org/opportunities/2011/07/19/is-you-car-killing-you-with-benzene.

Geisler, Norman L. *Christian Ethics*. Grand Rapids: Baker, 2010.

Glasser, Harold. "Naess's Deep Ecology Approach and Environmental Policy." *Inquiry* (39) 157–87.

Goldfarb, Theodore D., ed. *Taking Sides: Clashing Views on Controversial Environmental Issues*. 8th ed. Connecticut, MS: Duskin/McGraw-Hill, 1999.

Graham, Brian. "EPB 433-Health and Environmental Effects of Burning Waste Plastics." http://www.saskh20.ca/PDF/EPB433.pdf.

Greenway, Roger S. "The Cultural Mandate." In *Evangelical Dictionary of World Mission*, edited by Scott A. Moreau, 21–39. Grand Rapids: Baker, 2000.

Grudem, Wayne. *Systematic Theology*. Grand Rapids: Zondervan, 2000.

Hinrichsen, Don. "Stratospheric Maintenance: Fixing the Ozone Hole is a Work in Progress." *The Amicus Journal* 18.3 (Fall 1996) 35–39.

Kale-Dery, Severious. "China Willing to Support Ghana Resolve 'Galamsey' Challenges." *Daily Graphic*, May 30, 2018. https://www.graphic.com.gh/news/general-news/china-willing-to-support-ghana-resolve-galamsey-challenges.html.

Kyei, Yaw. "President Slams Foreign Firms . . . for Deplorable State of Mining Communities." *Ghanaian Times*, May 30, 2018. http://apanews.net/index.php/en/news/press-focuses-on-presidents-criticism-of-mining-firms.

Majeed, Yakubu Aboul. "'Green Republic Project' to Combat Climate Change Takes Off." *Daily Graphic*, June 1, 2018. http://thegreenrepublicprojectgh.org/press/

Mante, Joseph O. Y. *Africa: Theological and Philosophical Roots of Our Ecological Crisis*. Accra, Ghana: SonLife, 2004.

McFague Sallie. *Models of God: Theology of an Ecological, Nuclear Age*. Philadelphia: Fortress, 1988.

Mikkelson, David. "Do You Know the Danger of Turning on the A/C After Starting the Engine?" http://www.snopes.com/fact-check/air-scare, 2009.

Mintah, Ken Osei. "Norwegian Prime Minister in Ghana for 2 day visit." *Daily Graphic*, June 1, 2018, 16-17. https://www.myjoyonline.com/news/2019/May-31st/norwegian-prime-minister-in-ghana-for-2-day-visit.php.

Nissani, Moti. "Greenhouse Effect: An Interdisciplinary Perspective." *Population and Environment* 16.3 (July 1996) 109–37.

Opera News. "Galamsey Queen's Partner Missing." https://www.myjoyonline.com/news/2019/April-18th/jailing-chinese-galamsey-queen-wont-solve-ghanas-problems-snr-minister.php.

Owusu, Francis Osei. "Environmental Degradation an Issue of Christian Concern: A Case Study of the Impact of Illegal Mining Operations at Pokukrom. Master's Thesis, Trinity Theological Seminary Legon, 2016.

Poore, Patricia, and Bill O'Donnell. "Ozone." *Garbage* (Sept/Oct 1993) 15–18.

Reilly, Bernard J. "Stop Superfund Waste." *Issues in Science and Technology* 9 (Spring, 1999) 57–64.

Robertson, Roland. "Glocalization: Time-Space and Homogeneity-Heterogeneity." In *Global Modernities*, edited by M. Featherstone et al., 27–44. London: Sage, 1995.

Schonberg, David. *Ecology and Beyond: A Biblical Approach to Life, Relationship, and Environment*. Alexandria, MN: Caravan, 1995.

Simkins, Roland A. "The Bible and Anthropocentrism: Putting Humans in their Place." In *Reflecting God's Glory Together: Diversity in Evangelical Mission*, edited by Scott Moreau and Beth Snodderly, 18–23. Pasadena, CA: William Carey Library, 2011.

Trayner, Ian. "Climate Change, Brussels." http://www.guradian.uk/environment/oct30/eu-climatechange-funding-deal.

UTV News. "KMA Inspects compost Facility in Kumasi." Vision One News. "Densu River is Dead." Midday Audio News Item on Radio. https://www.ghanaweb.com/GhanaHomePage/NewsArchive/Man-22-drowns-in-Densu-River-737382.

World Health Organization. "Dioxins and their Effects on Human Health." http://www.WHO.int/en/news-room/fact-sheets/detail/dioxins-and-their-effects-on-human-health, 2016.

Fulani Herdsmen Traditions and Care for the Land

Haruna Y. Mogtari
Centre for Research and Islamic Studies, Akropong

Abstract

The paper exposes some of the significant practices of the Fulani people in relation to land care. It examines some traditions of the Fulani herdsmen and the use of the land in Ghana, which are of great importance. It provides a brief historical background to the Fulani in Ghana. Additionally, the paper discusses an important Fulani worldview termed as Pulaaku, *found in almost every Fulani group, and investigates their traditional knowledge on the use of land resources and its maintenance.*

Introduction

THE FULBE, WHO ARE popularly known as Fulani in West Africa, began to dominate the media waves during the 1990s as a result of the farmer-herder conflicts.[1] Since then these conflicts have continued to increase in number as herdsmen migrate to new grazing areas in farming communities across the country, as is the practice. Up until the last couple of decades of these conflicts, not many residents living in southern Ghana had negative perceptions of the Fulani even though they had been living in the north for the last century as hired herdsmen. Many Ghanaians will admit that they have

---

1. Tonah, *Fulani in Ghana*, 3-4.

for several decades been beneficiaries of Fulani pastoral services. Sad to say, the media, when reporting some of the farmer-herder conflicts, is biased against the Fulani. Such distorted images are a contributing factor to the prejudice, stereotyping, stigmatization, and marginalization of the Fulani. However, some aspects of Fulani mission consciousness and spirituality in Ghana can reveal how they approach land care. The issue is: can anything good come from the Fulani on land care in spite of all the negative imagery? This paper explores some of the pastoral practices of Fulani herdsmen that are relevant for the care of the land, trees, and nature.

The Fulani Imagery in Ghana

The Fulani belong to an ethnic group that refers to themselves as *Fulbe* (plural) or *Pulo* (singular), and they speak Fulfulde as their mother tongue. *Fulfulde* is part of the Niger-Congo family of languages. Palmer and Johnston reveal that the words *Fulani, Peul*, and *Fellata* are corrupted versions of *Fulbe* or *Pulo*.[2] In fact, the origin of the Fulani has led to numerous speculations, and their traditional accounts of themselves have also left different degrees of uncertainties about their origins. Some of the speculations have racial undertones which are less resourceful and problematic.

In 1949, in his study of Fulfulde, Joseph Greenberg suggested that the Fulani came from West Africa. Then just around the same period in the 1950s, Derrick Stenning discovered some reliable sources written by Arab and European explorers in AD 872 which further point to Western Sudan as the original home of the Fulani. Later, Stenning suggested that the Fulani mass emigration began from Western Sudan, precisely Senegal, spreading across through to Central Africa; for them the Fulani are Africans or partly Africans.[3] Delafosse suggests that Fulani emigrated for the first time from the kingdom of Tekrur in Senegal in the eleventh century AD, and by the beginning of the thirteenth century the Fulani built settlements and scattered throughout the sub-region.[4] By 1911, Fulani were already living in the Northern Territories of the then-Gold Coast. Their population, from an insignificant number, rose gradually through the first quarter of the twentieth century to about 300,000 in the 2000 Ghana population census.[5] It is actually difficult to know the real population of Fulani living in Ghana today since many of them live in inaccessible areas

---

2. Palmer and Johnston, "'Fulas' and Their Language," 121, 128.
3. Stenning, *Savannah Nomads*, 20.
4. Stenning, *Savannah Nomads*, 21.
5. Tonah, *Fulani in Ghana*, 2–3.

in the bush, and the fact that they are highly mobile. Their numbers are often grossly underestimated. For instance, the 2010 population census presents the Fulani total population as only 4,243.[6]

One can say that not all Fulani residents in Ghana are foreigners, though a reasonable number claim to be foreign citizenry. The Citizenship Act, 2000 (Acts 591), section 1-7 of Ghana's constitution, affirms descendants of early Fulani migrants (before and after 1957) who either by birth or naturalization and registration have since made Ghana their home as citizens. Thus, the second and third generations of Fulani descendants of the early migrants are legally Ghanaians by virtue of the fact that their parents and grandparents were born in Ghana or have naturalized or registered through marriage. This class of Fulani live close to or in local communities as hired herdsmen for indigenes and local authorities, and have been accepted as members of their society. It is a known fact that Fulani are not only working as herdsmen; some have diversified into the formal and informal sectors of the economy as doctors, nurses, teachers, politicians, drivers, traders, tailors, butchers, and Islamic scholars.

Apart from the "Ghana Fulani," there are also foreign herdsmen who seasonally migrate into the country to graze their cattle, trade in the cattle market, or find employment as hired herdsmen. Among the foreign herdsmen, there are two distinct groups—those who obtain permits from local authorities to settle on their land as tenants based on contractual agreements, and those who come into the country with their cattle and settle in the bush without any permission. Most often, it is these unlicensed foreign herdsmen, known as alien herdsmen, who illegally enter the country with their cattle and engage in grazing malpractices and alleged criminal activities leading to conflict with local farmers in some parts of Ghana. Notwithstanding, the Ghana Fulani herdsmen can sometimes be found culpable of grazing malpractice.

Instances of grazing malpractice can be cited here. In 2000, herders in the Volta Basin allowed their cattle to roam freely in the dry season to feed on their own. This uncontrolled grazing coincided with the time when farmers had planted their crops in the fertile basin. Thus, farmers had to keep watch over their crops to prevent them from being destroyed by the cattle. These farmers ended up shooting and wounding or killing Fulani herdsmen or their cattle.[7] It is believed that some chiefs in Ghana contribute

---

6. Ghana Statistical Service Data Extract, *2010 Population and Housing Census*, Northern and Upper Regions.

7 *Daily Graphic*, 10 June 2000; *The Ghanaian Chronicles*, 7 August 2000.

to the problems created by the herdsmen. The testimony of the District Agric Officer in Atebubu speaks well to this issue when he said:

> The chiefs and landlords in the area are competing amongst themselves to host the Fulani, particularly the alien Fulani herders from Niger and Nigeria. Many of them (the chiefs) have been able to acquire large cattle herds through renting out land to the Fulani. The Fulani are also caring for their animals for them at no cost. So, you can see why most of the chiefs and landowners are competing with each other to get the Fulani on their land. It is those chiefs and landlords who have lost out in this tough competition to host the Fulani who then turn around to complain about the destructive activities of the Fulani and then try to instigate sections of the community against the Fulani.[8]

Some chiefs prefer leasing land to the herdsmen because of the huge and varied benefit they gain from the herding activities. In May 2011, *Africa Report Magazine* documented that Joyce Bamford-Addo, Ghana's Speaker of Parliament, "announced the creation of a parliamentary committee to investigate conflicts between Fulani herdsmen and farmers."[9] Human rights activists in response to the announcement raised concern about "ethnic victimization," because the debate was dubbed the "threat and danger posed by the herdsmen."[10] The perennial farmer-herder conflicts have featured prominently in the media in recent times. In 2015, an incident was reported that the Ashanti Regional Security Council had staged a renewed operation captioned "Operation Cow Leg" within the Agogo community and its vicinity to control the activities of Fulani herdsmen. The aim was to bring peace and serenity to the area and to quench any confrontation that may arise between the residents of Agogo and the Fulani herdsmen. The inhabitants had constantly lamented about the pastoral activities in the area, which had had adverse effects on their farming businesses. Some armed Fulani guarding their cattle while they grazed were hostile to farmers and did not allow them access to their farms. Other complaints by famers in the Agogo area included contamination of water bodies, destruction of farm crops, and alleged raping of women by herdsmen.[11] In reaction, some farmers sprayed farmlands with herbicides or insecticides and burnt the grass with

---

8 Agricultural Extension Staff at Atebubu District, 2006.

9. Smith, "Parliamentary Watch," 37.

10. Smith, "Parliamentary Watch," 37.

11. Regional Security Council Report (2010), cited in Baidoo, "Farmer-Herder Conflict," 50.

the intention to keep cattle away, which ended up poisoning them.[12] In the Eastern Region, the *Daily Graphic*, under the heading "Fulanis Clash With Kwahus, Four Dead," reported that three Fulani herdsmen and an indigene had died in separate attacks.[13]

In Yaa Oppong's opinion, the common Ghanaian stereotype of the Fulani is that they are cattle raisers, cattle rustlers, and crop destroyers. She claims, however, that not all Fulani are pastoralists and dishonest people. She says the actions and inactions of a few Fulani may have influenced certain perceptions of the entire population. According to her, the "antinomadic" Ghanaian government policies and their direct interventions in certain parts of Ghana have further worsened Ghanaians' perceptions about the Fulani in the country.[14]

Fulani Herdsmen and the Care of the Natural Resources of the Environment

Fulani herdsmen have some traditional preventative measures for the maintenance of the environment. Their reason for applying these traditional precautionary methods is so that they can improve the growth of pastures to feed their cattle all year round. Indigenous Fulani abhor the felling of trees and the burning of grass. In the dry season, when grass is scarce, they will cut a few branches and not the entire tree to feed their cattle. The Fulani in Gushiegu shared their frustration about bush burning, which

> is killing everything in the environment; it kills the small animals and destroys everything–the grass and the trees, and our cattle are not able to graze. But when you use cutlass to cut a few branches of a tree to feed your animal in about a month time the tree will repair itself. On the other hand, if you burn the tree it will never repair but rather die perpetually.[15]

The effect of bush burning leading to loss of trees has had a huge negative impact on Fulani and their cattle, as well as other people who equally depend on the environment for survival. Depletion of the environment drastically reduces the fertility of their cattle, which in turn leads to rapid decrease in their herds. Additionally, when the grass is destroyed, they cannot build their houses and maintain the home because it is the grass and

12. Baidoo, "Farmer-Herder Conflict," 51, 52.
13. Afolley, "Fulanis Clash With Kwahus," para. 8.
14. Oppong, *Moving through and Passing on*, 1.
15. Fulani, in Gushiegu, Group Interview, 25 July 2015.

the trees they use to build their houses. Again, if the environment is depleted, their children will have to travel long distances for several weeks and months with their cattle to find pasture. For these reasons they usually do not encourage the destruction of trees and grass in the environment. As Harold Turner notes, in primal societies, man is

> akin to nature, a child of Mother Earth and brother to plants and animals which have their own . . . place in the universe . . . and may enter into totemic spiritual relationship with men in the way the environment is used; with . . . respect and reverence . . . without exploitation. Thus wanton destruction is often regarded as evil.[16]

The lives of pastoral Fulani, both economic and social, are dependent on their cattle and the environment. To the pastoral Fulani, owning a cattle ranch is their highest aspiration in life, which forms a significant aspect of one's existence and determines one's status in society.[17] So, Fulani herdsmen and their families can be seen in the rural communities with their cattle, living sometimes in the bush. Fulani have not only perpetuated cattle rearing for over a thousand years, but have improved the quality of their yield over the centuries.

The Fulani honor certain plants. In Busie, in the Upper West Region, the Fulani claimed they cut only parts of the *kankagalaa* and *golma* trees which are sacred to them to feed their cattle.[18] Similarly, in the Tumu districts, within the same region, the Fulani claim they only cut small parts of the *kahi* and *kakaligahi* trees because it is very dear to them.[19] Peter M. Braimah, a Pulo, and an ordained minister (retired) with the Assemblies of God Church testifies:

> There are some trees that we honor . . ., like shea nuts tree. We do not like to cut them. The second one is the *dawadawa* or *dausari* tree because it can feed man with its fruits. The *sakumzoro* tree, we do not also like to cut them because we eat the fruit. Any tree that bears fruit Fulani honour it, they do not destroy it.[20]

Just as the cocoa tree is important to the Akan, so is the Shea nut tree indispensable to the life of the people of Northern Ghana, including the

---

16. Turner, "Primal Religions of the World," 30.
17. Fulani, in Gushiegu, Group Interview, 25 July 2015.
18. Ard, Interview, 20 September, 2015.
19. Harun and Behiduma, Interview, 22 September 2015, Tumu.
20. Braimah, Interview, 25 July 2015, Saboba.

Fulani. *Dawadawa* is the favorite delicacy of the people of the Northern and Upper Regions, and the Fulani also use it in preparing their food.

In the Central Gonja District, herdsmen claim that they practice a shifting system of grazing, where cattle graze at a particular area at one time and then move to a new area in order not to completely deplete the grass, thus leaving it to replenish itself. Fulani also value water so much so that they can travel many kilometers for days and weeks in search of it for their herds. Some Fulani say that they do not encourage people to wash in or misuse water resources for the reason that these activities diminish the water quantity and quality.[21]

Notwithstanding, there are some herdsmen—especially the alien ones, who are known for their illegal activities—who fail to adhere to the Fulani traditional knowledge of preserving the environment and thereby allow their cattle to destroy farm crops and drink from water sources and thus pollute them. Ali Husein, a Pulo, has asserted that these alien herdsmen do not value the land, and neither do they think of preserving its resources. He further disclosed that they are ignorant about the risk associated with not preserving the land. Ironically, he reported that Fulani herdsmen in his area complain about their inability to acquire land from local chiefs due to the increasing number of herds and farms. He then advised that Fulani herdsmen should creatively find a new approach to graze their cattle, for instance in ranches, because a time will come when there will be absolutely no land space for their cattle to graze due to the proliferation of farms everywhere.[22]

A Launch into the Significance of *Pulaaku*

Since the subject of this study is Fulani herdsmen, it will not only be essential, but appropriate, even if briefly, to highlight some of the important and specific cultural elements that make them a distinct ethnic group. In this case, I will focus on *Pulaaku*, a dominant and common tradition found in varied expressions among different Fulani groupings scattered across West Africa which appear to permeate their whole life.

In Ghana, within the Fulani society lies a mine deposit of great cultural diversity hidden from persons who do not belong there. *Pulaaku* is one of the most significant aspects of Fulani worldview; their vital life experiences revolve around it—from their daily activities to fulfilling the customs and traditions of the society. *Pulaaku* implies Fulani society or "the essence or comportment of . . . a Pulo, which in its ideal state consists of

21. Fulani, in Buipe, Group Interview, 1 August 2015.
22. Hussein, Interview, 18 July 2015.

patience . . . fortitude (*munyal*), wisdom (*hakkillo*) . . . modesty and reserve (*semteende*)."[23] A Pulo's ability to speak "in" Fulfulde is a key feature of *pulaaku*. Fulfulde as an indispensable cultural element means not just a spoken language, but a whole range of rights and duties[24] distinctive to a Fulani. *Pulaaku* covers a wide range of wisdom and knowledge that makes it absolutely impossible for anybody to attempt to learn everything about it.

A Pulo who possesses *pulaaku* will not behave in a manner that does not befit a "true Fulani."[25] For instance, he or she will not dress inappropriately, in a way that contradicts the Fulani dress code, that is to say the wearing of long gowns that covers most part of their body. They also abhor bad eating manners, breaking of taboos, and defecating in the open. Avoiding shame or fear of shame features prominently in what it means to possess more *pulaaku*.[26] Similarly, a Pulo who possesses *pulaaku* is expected to be kind, humble, peaceful, patient, gentle, respectful, and generous. It therefore means that any Pulo who engages in contrary behavior such as violence or some kind of deviant behavior has fallen out with the *Pulaaku* and should be put to shame or punished.

Fulani tradition has it that if a Pulo who possesses *pulaaku* enters a house and a mat is spread out for him or her to sit on, the one is expected to sit at the edge and not the middle part.[27] Here, the moral lesson is that if he or she sits in the middle then he or she has not learned to be modest.[28] Thus, humility is a great virtue to Fulani. As noted earlier, there is a characteristic attitude in the rearing of cattle which is a distinctive attribute of Fulani in West Africa. Comparatively, some of the character traits of the Fulani are similar to that of the Masai of Kenya in East Africa, who are noted for their hospitality, generosity, gentleness, piety, and association with cattle.[29]

Baliigo/Walitaago (Compassion)

This section focuses on an aspect of *Pulaaku* associated with rearing of cattle, known as *baliigo* or *walitaago*, found among some Fulani clans in Ghana. The same practice is popularly termed as *habanii* among herdsmen of other

23. Smith, *Burkina Faso Fulfulde-English*, 180.
24. Stenning, *Savannah Nomads*, 55.
25. Tonah, *Fulani in Ghana*, 133.
26. Oppong, *Moving through and Passing on*, 151–52.
27. Sumaila Mohammad Gedi, Interview, 29 September 2015, Bawku.
28. Fulani Chief (In the company of Bari Harun & Bari Behiduma), interview, 22 September 2015, Tumu.
29. Donovan, *Christianity Rediscovered*, 50.

West African countries. The words *baliigo* or *walitaago* may be defined as "to be helped"[30] or "to receive help."[31] In other words, *baliigo*, *walitaago*, and *habanii* connote an act of compassion shown to a Pulo by a kinsman or friend when something unfortunate happens to him, or especially when he becomes a victim of cattle rustling. It is an embodiment of comradeship and loyalty shared by Fulani herdsmen. Bashiru bin Umar, describes how *habanii* is perfomed:

> A Pulo who tends animals, may enter into an agreement with a colleague. He may give a female cow to the associate and when it gives birth, either male or female, it is a gift for the partner. When that same female cow, given to him earlier, gives birth a second time, the colleague will have to send back the cow and the new heifer to the owner. This practice is called *habanii*, and the person who offered this favour will become one [bonded in relationship] with him who received. If the person who offered the cow is not even a Fulani it is still *habanii*.[32]

The expected outcome of *habanii* is peace, love, solidarity, and a lasting relationship which ultimately deepens the family bond of the parties involved. It is clear from the above that *habanii* can also take place between Fulani and non-Fulani because it is not discriminatory. Interestingly, it is such acts of compassion that inform one's relationship with vegetation.

Reflections on the Herder-Host Conflicts through the Eyes of Scripture

According to Kwame Bediako, although the gentile Corinthians and Galatians had cultural backgrounds different from those of the Jews, Paul quoted the Old Testament and related it to these gentiles. For him, the relationship between the gentiles' context and that of the Jews was authentic, and the Old Testament was relevant for the former because it communicated to them through adoption into the new Israel.[33] There seems to be some similarity between the Fulani herdsman and the occupation and pastoral practices in the Bible.

As we earlier noted, there have been perennial conflicts between Fulani herdsmen and farmers, who compete for scarce water and land

---

30 Smith, *Burkina Faso Fulfulde-English*, 84.

31 Smith, *Burkina Faso Fulfulde-English*, 84.

32 Bashiru Bin Umar, Interview, 11 October, 2015, Accra.

33. Bediako, "Scripture as the Hermeneutic," 3.

resources in some parts of Ghana, mainly because of climate change. These farmer-herder conflicts are not new to the world, as we can read stories of similar conflicts between some patriarchs of the Old Testament and their host communities. There was an instance of two patriarchs fighting among themselves over pasture lands. In Genesis 13, we read about the cause of separation of Abram and Lot, his nephew:

> Abram had become very wealthy in livestock . . . Now Lot, who was moving about with Abram, also had flocks and herds . . . But the land could not support them while they stayed together, for their possessions were so great that they were not able to stay together. And quarrelling arose between Abram's herdsmen and the herdsmen of Lot . . . So Abraham said to Lot, 'Let's not have any quarrelling between you and me, or between your herdsmen and mine, for we are brothers . . . Let's part company. If you go to the left, I'll go to the right; if you go to the right, I'll go to the left . . . The two men parted company (Gen 13:2-11).

The cause of the conflict between Abram and Lot was mainly because of shortage of water and land resources. These causes are no doubt similar to those of the conflicts between the Fulani and their hosts in Ghana. Abram and Lot dialogued and came to the consensus to part ways.

Bruce K. Waltke and Cathi Fredricks note that Abram recognized his nephew as an equal and valued the peaceful relationship between them more than his personal wealth. It indicates that at times brothers have to separate to find peace since "a major theological concern of this story is the priority of peace between brothers."[34] However, we also observe Abram's humility and generosity in his approach to resolving the conflict, which he did through dialogue and choosing to let go of his rights. By implication, the initial step to resolving the conflict situation in Ghana is for host communities of herdsmen to learn to be tolerant, especially in the event of farm destruction because retaliation never leads to any form of peace. Nevertheless, it does not also mean that the fundamental human right of farmers and even herdsmen should be trampled upon. It has become pertinent that the government institute the right kind of laws and ensure that these regulate farming and grazing activities to protect the opposing sides.

Moses revealed Israel's social responsibility toward strangers who live in and around them: "Do not mistreat an alien or oppress him, for you were aliens in Egypt" (Exod 22:21). God reminds Israel about the ordeal they experienced in the hands of their overlords, the Egyptians, as strangers, when they dwelt among them. Unlike the Egyptians, God shows his children a better

---

34. Waltke and Fredricks, *Genesis*, 222–24.

way to handle vulnerable foreigners, that is, with dignity and kindness. This social concern matters so greatly to God that he recalls: "The alien living with you must be treated as one of your native-born. Love him as yourself, for you were aliens in Egypt" (Lev 19:34). In Luke, Jesus' response to the question of the lawyer implied that a neighbor could be a total stranger, and that utmost compassion and love should be shown to him (Luke 10:27–37). In doing this, one is not only fulfilling the second, weightier commandment in Scripture (Matt 22:38–39), but also becomes a witness for the faith. In our context, God has called the church in Ghana to show the same social obligation and pastoral care toward Fulani herdsmen.

Waltke and Fredricks further make a profound statement: "Indeed, peace-making and reconciliation are so central to God's character as revealed in Christ."[35] Perhaps, that is why similar stories of herder-host conflicts are told in Genesis 21 and 26, but in these cases the conflicts ensued between Abraham and Abimelech, and then again Isaac and the latter. In the first case, Abraham made a complaint to Abimelech:

> about a well of water that Abimelech's servants had seized. Abimelech said, 'I don't know who has done this. You did not tell me, and I heard about it only today . . . So Abraham brought sheep and cattle and gave them to Abimelech, and the two men made a treaty. Abraham set apart seven ewe lambs from the flock, and . . . Abraham replied, accept these seven lambs from my hand as a witness that I dug this well . . . And Abraham stayed in the land of the Philistines for a long time (Gen 21:25-34).

According to Waltke and Fredricks, the main issue at stake was about grazing and well rights that were essential for Abraham and his livestock. They remark that Abraham's complaint to Abimelech suggests a resolve to find the right solution to the conflict situation between him and Abimelech's servants. Abimelech readily compromised with Abraham to ensure that there was no violence. Abraham demonstrated his determination to settle the dispute without violence by giving gifts to Abimelech, who in turn accepted the gifts and swore an oath before witnesses as evidence of his willingness to stop the conflict. Abraham stayed in the land of the Philistines for a long time as a sign of the positive outcome of the treaty.[36] In Genesis 26, the writer says that:

> Isaac planted crops in that land . . . he had so many flocks and herds . . . that the Philistines envied him. So all the wells that his

---

35   Waltke and Fredricks, *Genesis*, 222–24.
36   Waltke and Fredricks, *Genesis*, 299–300.

father's servants had dug . . . the Philistines stopped up, filling them with earth . . . Isaac's servants dug in the valley and discovered a well of fresh water there. But the herdsmen of Gerar quarrelled with Isaac's herdsmen and . . . Meanwhile, Abimelech . . . answered, 'We saw clearly that the LORD was with you; so we said, 'There ought to be a sworn agreement between us'—between us and you. Let us make a treaty with you that you will do us no harm, just as we did not molest you but always treated you well and sent you away in peace . . . Isaac then made a feast for them, and they ate and drank. Early the next morning the men swore an oath to each other. Then Isaac sent them on their way, and they left him in peace. (Gen 26:12-31).

First, Waltke and Fredricks suggest that when Isaac's wealth in herds and flocks increased, there was the need for them to dig enough wells to meet the demand for water. But this led to envy by the Philistines, and they struggled for the wells Isaac's herdsmen had dug. Second, the Philistines broke the peace agreement Isaac's father, Abraham, had earlier made with them (Gen 21:28). Once again Isaac entered into another peace treaty with the Philistines.[37] Waltke and Fredricks further suggest that it is Isaac's closer walk with God and obedience to him that led to his peace with the Philistines.[38]

In the three herder-host conflict cases presented above, what stands out is the judicious use of the land, water, and protection of animals. Aspects of *pulaaku*, and especially *habanii*, can be seen in the pastoral activities. Virtues such as kindness, humility, patience, gentleness, and generosity are all essential in promoting peace and land care. Anyone who does not act in ways that promote land care has fallen out with *pulaaku* and should be put to shame. These leaders in the Bible took bold steps to act as mediators, to establish and constantly renew their agreement concerning rights of grazing and water resources.

Conclusion

It has been shown from the foregoing discussion that it is not entirely truthful to accuse the Fulani of not caring about the environment. Their tradition indicates vast resources and mechanisms within it that, if adhered to, can become a great asset for maintenance of the environment. A careful reflection on the herder-host conflicts in Genesis 21 and 26 about the patriarchs, especially Abraham, Isaac, and their host nations, also teaches us ways to

---

37 Waltke and Fredricks, *Genesis*, 369–71.
38 Waltke and Fredricks, *Genesis*, 371.

resolve the perennial conflicts between farmers and herdsmen in Ghana and elsewhere in West Africa. It is only when the gospel of peace reaches the hearts of herdsmen and farmers that true reconciliation begins; first with God, then humanity, and the environment.

Bibliography

Abugri, George Sydney. "Agogo: Following the Trails of the Armed Fulani Nomad." *Daily Graphic*, February 6, 2016. https://www.graphic.com.gh/features/opinion/agogo-following-the-trail-of-the-armed-fulani-nomad.htm.

Afolley, George. "Fulanis Clash With Kwahus: Four Death." Daily Graphic, January 28, 2015. https://www.graphic.com.gh/news/general-news/fulanis-clash-with-kwahus.html.

Afolley, George, et al. "Fulani Herdsmen in Trouble: Police, Soldiers to Flush Them Out of Agogo, Fear Grips Begoro Residents." *Daily Graphic*, February 4, 2016. https://www.graphic.com.gh/news/general-news/fulani-herdsmen-in-trouble-as-police-soldiers-to-flush-them-out-of-agogo.html.

Asiedu, Abednego Asante. "Fulani Herdsmen More Dangerous than Boko Haram—DCE fears." https://www.myjoyonline.com/news/2016/February-4th/fulani-herdsmen-more-dangerous-than-boko-haram-dce-fears.php.

Baidoo, Ibrahim. "Farmer-Herder Conflict: A Case Study of Conflict between Fulani Herdsmen and Farmers in the Agogo Traditional Area of the Ashanti Region." MA diss., University of Ghana, Legon, 2014.

Barclay, William. *The Daily Study Bible; Letters to Timothy, Titus and Philemon*. Edinburgh: Saint Andrew, 1956.

Bediako, Kwame. "Scripture as the Hermeneutic of Culture and Tradition." *Journal of African Christian Thought* 4.1 (June 2001) 2–11.

Boateng, Caroline. "Establish Land Banks to Deal With Menace of Fulani Herdsmen." *Daily Graphic*, February 18 2016. http://www.graphic.com.gh/news/general-news/58587.

Brueggemann, Walter. *Genesis*. Interpretation. Atlanta: John Knox, 1982.

———. "Flush Out Fulani from Our Land." *Daily Graphic*, 3 February 2016. http://www.graphic.com.gh/news/general-news/57598.

Donkor, Kwadwo Baffoe. "'Operation Cow Leg' Reactivated at Agogo." *The Daily Graphic*, January 16, 2015. https://www.graphic.com.gh/news/general-news/operation-cow-leg-reactivated-at-agogo.html.

Donovan, Vincent J. *Christianity Rediscovered: An Epistle from the Masai*. Maryknoll, NY: Orbis, 1982.

Oppong, Yaa P. A. *Moving through and Passing on: Fulani Mobility, Survival, and Identity in Ghana*. New Brunswick, NJ: Transaction, 2002.

Palmer, H. Richmond and Harry H. Johnston. "The 'Fulas' and Their Language." *Journal of the Royal African Society* 22.86 (1923) 121–30.

Smith, Richard W., ed. *Burkina Faso Fulfulde-English/English-Fulfulde Dictionary*. 2nd ed. Tenkodogo-Ouagadougou, Burkina Faso: SIM Burkina Faso, 2007.

Stenning, Derrick J. *Savannah Nomads: A Study of the Wodaabe Pastoral Fulani of Western Bornu Province Northern Region, Nigeria*. Hamburg: Lit, 1994.

Tawiah, Ohemeng. "DCE Begs IGP for Weapons to Deal With Nomadic Herdsmen Crisis." Ghana|Myjoyonline.com, 9 March 2016. https://www.myjoyonline.com/news/2016/March-9th/police-beg-igp-for-weapons-to-deal-with-agogo-crisis.php.

Tonah, Steve. *Fulani in Ghana: Migration History, Integration and Resistance.* Accra: University of Ghana Press, 2005.

Turner, Harold. "The Primal Religions of the World and their Study." In *Australian Essays in World Religions,* edited by Victor Hayes, 1–15. Bedford Park, SA: Australian Association for the Study of Religions, 1977.

Waltke, Bruce K., and Cathi J. Fredricks. *Genesis: A Commentary.* Grand Rapids: Zondervan, 2001.

Desacralization of Gold in Southwest Burkina Faso: A Christian Response to Gold Mining and its Consequences on Creation

Ini Dorcas Dah

Akrofi Christaller Institute of Theology,
Mission & Culture, Akropong

Abstract

This paper aims to reflect biblically and theologically on some observed ecological consequences of disregarding the sacredness of gold in Lobi culture, and the resultant devastating gold mining in Southwest Burkina Faso. It is argued that issues affecting the ecology in Ghana are similar to those of their neighbors in Burkina Faso. The hope is to provide a Christian educative response that may motivate attitudinal change, especially among the youth, toward a religious and moral approach to gold mining and keeping ecological integrity.

Introduction

FRENCH MONOGRAPHS OF 1902 provide the early documentations of the discovery of gold in Lobi land, a town in Southwest Burkina Faso, and indicate that only women engaged in occasional gold mining, and at a small-scale level.[1] The women exchanged the gold for axes, iron, salt, and other

1. Archives Nationales de Côte d'Ivoire, Abidjan (ANCI-Abidjan), Monographie de Bondoukou, Colonie de la Côte d'Ivoire, Haute Côte d'Ivoire Orientale (deuxième

useful commodities.² The Lobi people neither engaged in large-scale gold mining, nor wore gold trinkets until the 1990s, because they believed gold had power to swallow people's souls. ³ In their understanding, God had placed gold inside the rocks on the hills so that no one could access it in large quantities. To break the rocks in search for gold was to bring untimely death upon oneself.⁴ Therefore, even when the Lobi found a rock containing a large gold nugget, they would not mine it, but instead would take it to the top of the hills and leave it there as a sacred stone.⁵

At present, things are different. Ken Gnanakan indicates, "Greed and exploitation have been around ever since the fall of humanity and have surfaced globally in recent times."⁶ In recent times, large numbers of people from different parts of the world have been attracted by and drawn, perhaps by greed, into Lobi land for gold exploitation through small-scale mining. The wanton activities of these foreign miners in Lobi land have influenced the indigenous youth to change their ecocultural dispositions and attitudes, which are affecting the ecosystem both physically and spiritually. Kwame Bediako is right when he argues that "Religious and moral values, things deemed acceptable and unacceptable, are all elements of culture. They affect how particular persons and their social groupings understand themselves, how they understand the world, how they think or relate to their natural or physical environment."⁷ This implies that a person's culture influences the way that person thinks. My observation is that disdaining the cultural belief that gold is sacred among the Lobi has led to wanton excavation of large tracts of land in search of gold, first by foreigners, and later, the indigenes, especially the younger generation.⁸ The problem associated with the land degradation in Burkina Faso is no different from what is happening to their neighbors in Ghana, and there is little attention paid to the repercussions of miners and artisanal mining and disregarding the indigenous culture and perceived sacredness accorded to gold. I propose that the lack of educative emphasis on the religious and moral implications of gold mining in Lobi

Cahier, troisième cahier et quatrième cahier), Les Etats de Bouna (Suite) Par le Lieutenant Chaudron du 1er Régiment de Tirailleurs Senegalais-Commandant la Circonscription de Bouna.

2. ANCI-Abidjan, Monographie de Bondoukou, Colonie de la Côte d'Ivoire, Haute Côte d'Ivoire Orientale (deuxième Cahier, troisième cahier et quatrième cahier).

3 Cécile de Rouville, *Organisation Sociale des Lobi*, 42.

4. Schneider, "Extraction et traitement ritual de l'or"191.

5. Schneider, "Extraction et traitement ritual de l'or," 191

6. Gnanakan, *God's World*, 6.

7. Bediako "Gospel and Culture,") 8.

8. Observation, 13 April 2017, Djikando (Gaoua).

land today contributes to this wanton engagement of the many youth in destructive gold mining activities. I follow Ebenezer Blasu's argument from his studies among the Sokpoe-Eʋe in Ghana and Ogbu Kalu in Nigeria that the absence of any visible retribution of humans who violate ecorules of deities and ancestral spirits demystifies the deities and nullifies the fear impulsion for creation care. For Blasu, "Many community members may become unwilling to cooperate on locally initiated ecoprojects, because the impulsion factor—fear of the deities—is no longer there."[9] This is the current case in Southwest Burkina Faso. Because the indigenous youth did not see any spiritual or physical punishment on the foreign miners, they also engaged in it. The situation is becoming worse and the numbers keep increasing because the first indigenous young people to undertake gold mining are not also facing any negative effects from the deities.

Consequences of Gold Mining in Southwest Burkina Faso

There are numerous consequences related to gold mining in Southwest Burkina Faso. However, for the purposes of this paper I will focus only on issues related to the environment, economy, society (hygiene), and education.

Environmental Consequences

In Southwest Burkina Faso, hectares of farms have been destroyed through gold mining by excavating the soil. The massive excavation of land in the area is causing serious environmental damage, particularly rendering agricultural soil and activity unproductive, worsening the helpless livelihood situations of many subsistence farmers.[10] In my view, the situation of making arable land unproductive through burying the humus of top soil by uncontrolled mining is unhelpful in attaining goal 2 of the UN 2030 Agenda, which is to "End hunger, achieve food security and improved nutrition and promote sustainable agriculture."[11] The promotion of sustainable agriculture and any related activity that can contribute to the well-being of the world population can only be possible if there is sufficient available farmland for people to undertake agricultural activities. It is thus difficult

---

9. Blasu, "Christian Higher Education as Holistic," 93. See also Kalu, "Sacred Egg," 241.

10. Tapsoba, "Batié: L'or, le métal qui divise!', Mutations No 15 du 15 octobre 2012. Bimensuel Burkinabè paraissant le 1er et le 15 du mois http://lefaso.net/spip.php?article51354.

11. United Nations, "Transforming Our World," para. 4.

to think of a better food supply in Africa if huge farmlands are excavated by gold miners, such as occurs currently in Southwest Burkina Faso.

Related to the challenge of destroying the humus in top soil, another significant effect of gold mining on agriculture is the massive removal of trees. I have personally witnessed the cutting down of trees in the process of creating access roads to mining sites. This affects rainfall patterns and intensity. Because of the excavation of the soil, it becomes difficult for new plants to grow in such areas. This means that people are not able to undertake their necessary agricultural activities.

Mining activities in Djikando have left open holes in the ground and scattered boulders all over the village. Some of the holes are located one yard away from some houses. No farming activity is successful in the village anymore, because of the destruction of the land. And the villagers, especially the youth, have also joined the foreigners to engage in illegal mining.

In addition to land excavation and all the consequences affiliated with it, the water bodies in the area are being destroyed. For instance, the river separating Gaoua and Djikando is completely colored and dangerous, not only for human life, but also for the fish in the water. Growing up in Gaoua, I knew that Djikando people used to supply the local population with fish from that river. Today all fishing activities have almost come to a standstill in the area. Gnanakan indicates that "Animals in the wild must be recognized as having certain needs for maintaining their life, their 'creatureliness' as willed by their Creator, their habitats, and their kinds. Destroying the animal world upsets not only the animals but also the ecological balance."[12] Gnanakan's statement implies that threatening animals in their biotope is another way of harming them and also offending God. The case of Southwest Burkina Faso is thus a concrete example of human beings' actions contributing to not only harming wildlife, but also hurting God, who created those animals.

My interactions with the people suggest that most of the youth are unfamiliar with biblical concerns about the environment. For instance, it is true that in Genesis 1:28 "God said to them [Adam and Eve], 'Be fruitful and multiply and fill the earth and subdue it and have dominion over the fish of the sea and over the birds of heavens and over every living thing that moves on the earth.'" Yet, when God gives us dominion over the universe, we do not own it but are only stewards over it. God allows us to use the elements, but a misuse of them is "ecological sin" before God, according to Ecumenical Patriarch Batholemeo I of the Greek Orthodox church.[13] It is clear from the

---

12. Gnanakan, *God's World*, 204.
13. Ecumenical Patriarch, Bartholomew I. Cf. Chryssavgis, "Address in Santa

Scriptures that instead of destroying it, God had specifically instructed that we till the ground and also keep his garden (Gen 2:15).

Barnabe Assohoto and Samuel Ngewa, commenting on Genesis 2:15, note that God is the One who created the world, but holds humanity responsible for maintaining what has been created.[14] Thus, in both Genesis 1:28 and 2:15, humanity are called to be keepers of God's creation rather than greedy users of it because God will hold human beings accountable for the misuse of creation's components.

Economic

Economically, gold mining is profitable to some individuals in the country and in the communities in which it is being practiced. For instance, one of the latest and biggest hotels built in Gaoua is owned by a person who is highly involved in gold mining and who encourages people to give out their lands for mining. Some people praise him for contributing to the development of the town by building such an infrastructure, which is uncommon in that entire area of the country. However, even though he is praised for his achievement in building such a hotel, it is evident that this was done at the expense of the lives of other people in the area because, as noted above, some villages are being completely destroyed and all other aspects of life are also negatively affected by the repercussion of gold mining. He is probably praised for his achievement because of the impressive infrastructure and also the fact that most people see development only from the material/economic perspective while forgetting about the social, environmental, and spiritual sides. This type of development is incomplete because it does not consider the lives of human beings in the area holistically.

Darrell Whiteman suggests a wheel for measuring integral human development as comprising "the spiritual growth, the personal growth, the social growth and the material growth."[15] According to Whiteman, one can only talk about development when there is a balance between the four aspects that form the wheel.[16] Thus this hotel as a major infrastructure in Gaoua is only seen from the material growth aspect, forgetting about the three other dimensions, without which development is imbalanced and therefore not considered as development. Achievement of this kind is also not sustainable because it does not consider the impact on future generations, since

---

Barbara, California."

14. Assohoto and Ngewa, "Genesis," 11.
15. Whiteman, "Bible Translation," 126–27.
16. Whiteman, "Bible Translation," 126–27.

sustainable development, according to the Brundtland Commission's definition, is "development that meets the needs of the present without compromising the ability of future generations to meet their own needs."[17]

The 25th principle of the Rio Declaration also states that "Peace, development and environmental protection are interdependent and indivisible."[18] From both the Brundtland commission's definition and the Rio Declaration it is obvious that one cannot talk about development if human life is under threat either in the present or in the future. Thus, it is good to celebrate current achievements, but when what is being celebrated is directly done at the costs of not only the current but also the future generations' peace and security, then moral issues need to be put into perspective.

It is also evident that, economically, the local communities, and especially those who are portioning their land to sell to gold miners, do not get any sustainable income because what they are paid, most of the time, is not enough to help them improve their current social circumstances, let alone leave something for future generations. The danger is that many people sell their farmland to gold miners and after they use the money they become worse off than they were before.

Social Consequences

The presence of gold miners, both legal and illegal, has led to the increase of insecurity in the area. More than sixty cases of armed robbery were recorded in 2015 in a small locality of the Batié area. In spite of security forces there, more cases were recorded in 2016. The large-scale mining in the area is therefore a serious threat for both land and humans. In addition to this insecurity, gold mining is even a threat to family structures. There has been an increase in the division of families because of controversial opinions between family members over gold mining. There have been cases of imprisonment of people in some villages due to litigation over gold mining, with social enmity and interpersonal suspicion equally on the rise.

Prostitution is also on the rise in gold mining sites, particularly among young unmarried women and sometimes even married women, who go to sell food and wares at the mining sites, but end up having affairs with the miners in order to improve their financial conditions. For instance, in 2016, a young woman was wounded and robbed at one of the mining sites in the area because three men had sexual intercourse with her

17. Integrating Environment and Development, 10.
18. Lehtonen et al., *Le Développement Durable*, 11.

consecutively after she collected CFAF15,000 from them.[19] There is also a general belief that having sex in the area attracts gold, and so this further fuel sexual activity at the sites.

However, Colossians 3:5–6 (ESV) reads, "Put to death therefore what is earthly in you: sexual immorality, impurity, passion, evil desire, and covetousness, which is idolatry. On account of these the wrath of God is coming." In these verses, Paul not only condemns sexual immorality and other related evil practices that go on in societies, but greed is also seen as idolatry. In the French translation (Nouvelle Version Louis Second Révisée), verse 6 of this text reads, "C'est pour cela que vient la colère de Dieu sur les rebelles" (That is why God's wrath falls on the rebels).[20] From a biblical perspective, this is an indication that by getting involved in sexual immorality, compounded with greed as the backing force behind gold mining in Southwest Burkina Faso, human beings place themselves as rebels against God and consequently incur God's punishment.

Immorality is not only a defilement of the people involved, who are created in the image of God and are also to be seen as temples of the Holy Spirit (1 Cor 3:16–17; 6:12–13), but it also defiles the land in general. Among the Birifor, it is sacrilegious to have sexual intercourse in the bush, and adultery is abhorred, especially for women.[21] Thus, the desacralization of gold in the area has led to the desecration of the land, thus violating the sacredness of human life. The effect of all these, therefore, is the devaluing of human life and land.

Another major issue affecting people at mining sites today is hygiene. Most of the mining sites I visited in the area have serious hygienic problems. Not only are the people dirty, but the air in the area is polluted with heavy clouds of dust.

Education

In some villages, children are abandoning school to go and work in mining sites, where they can get quick cash. However, this jeopardizes their future because most people have already sold their lands to gold miners and so these children will no longer have land to farm and live on as their forefathers used to. Such children will not be able to get jobs with the government due to their low levels of education. Neither will they be able to get jobs

---

19. The names of the people involved in this situation have been omitted in order to protect their identity.

20. Free translation by the author.

21. Dah, "Birifor Women Communicating the Gospel," 70–71.

with the government due to their low levels of education. The proximity of some mines to local schools is a real challenge to children's education. For instance, the mining site of Djikando is only about 100 meters from the primary school in the village.[22] This has led to many school children leaving the school to work at the mining site.

The Birifor initially did not like school because of their bad experiences there under French colonization, and thus perceived the introduction of schools as another method of the French colonial government to subdue them.[23] According to Madeleine Père, negative colonial actions and attitudes led the Birifor to "*bin nvɔr*" (put a mouth down: take an oath) because they and their descendants would never have any relationship with anything related to a white person, including formal education.[24] Writing about school in the 1920s, Maxime Compaoré notes (first in French, then in English):

> Enjoyer son enfant à L'école n'a pas toujours été une chose facile. La méthode de récrutement, l'éloignement des écoles, la séparation d'avec la famille, la réduction de la main-d'oeuvre familiale occasionnée par le départ des enfants, étaient autant d'éléments qui expliquaient la réticence des parents devant L'école. A ces raisons s'ajoute la méconnaissance de la finalité de L'école.[25]

> It has not always been an easy task for one to send one's child to school. The method of enrolment, the distance of schools, the separation of children from their families and the reduction of the family workforce through the departure of the children were enough reasons for the parents' reluctance in sending their children to school. The final reason was the uncertainty that surrounded one's future prospects after schooling.[26]

School was not only a nightmare for some parents in Burkina Faso, but the learning conditions in some villages also led to a lack of interest in encouraging children to attend school. According to Massofa Kambou, the head master of the primary school of Djikando:

> Gold mining has a negative impact on the school because even though government sends food stuff to the school, there are times there is no food for the children. Therefore, they

---

22. Observation, 20 April 2017, Djikando.
23. Dah, *Women Do More Work*, 81–83.
24. Père, *Villages en Transition*, 2:548.
25. Compaoré, "L'école au quotidien en Haute-Volta pendant," 364.
26. Free translation by writer.

> automatically think that going to the mining site to earn 1000f or 2000f at the end of the day is better than going to sit in a classroom for a whole day without food to eat.[27]

Kambou also explained that even some young men who are attending secondary schools in Gaoua town are tempted to come back to the village and join the miners. Those young men assume that they are suffering in town going through education even to the point of having no food to eat while other people from different areas of the country are making money from their own villages.

The presence of gold miners in the village of Djikando has therefore worsened the challenges concerning formal education in the area. Parents confronted with poverty would prefer that their children go to the mining sites in order to return home with cash to buy food, rather than going to school and starving for the whole day with no hope of even finding a meal at home in the evening. As a result, the number of pupils keeps decreasing in the primary schools.[28] To my question of what the government is doing to change the situation in the village of Djikando, which has been completely devastated by gold mining, particularly in the area of education, the headmaster responded:

> News have gone all over, even Plan Burkina Faso came here and noticed the situation. Some authorities have come here several times, but it is the same issues I indicated earlier. We cannot automatically blame the miners, but parents are also responsible in following up with their children whether they are really in school during the day. Even when a child does not come to school and we call her or his parents, they do not come. So the first people who are responsible for the lack of children's interest in the school are their parents and the mining activities are also contributing to their withdrawal from the school. I have even requested from the authorities that they should provide documents stating the distance between a school and a mining site because there is nothing we can do as teachers if we do not have any legal document about the distance between gold mining and school. All regional authorities are informed about the situation, but the measures to stop the mining at the doorstep of the school have not worked out.[29]

---

27. Massofa Kambou, Interview, 20 April 2017, Primary School of Djikando, Gaoua.

28. Kambou, interview, 20 April 2017.

29 Kambou, Interview, 20 April 2017.

The above words of the headmaster show how helpless the teachers are in this village because of the threat of gold mining toward their school. It also exhibits the level of social injustice that is linked to human greed, to the point that even the local authorities are not mindful of the future of the children, whether through their education or their safety in entering the holes in search for gold. Gold mining is, therefore, a serious challenge to the education of children in the village, leaving the headmaster and his team powerless in addition to the historically negative attitude toward formal education in the area. This also overpowers one of the goals of the 2030 Agenda, which is to "ensure inclusive and equitable quality education and promote lifelong learning opportunities for all."[30] Gold mining, therefore, has a doubly negative impact in Djikando and many other villages in the area.

The Church's Response

With all of the negative impacts of gold mining upon human life and creation, De Rouville's statement that the "Lobi would not wear gold because they culturally believed that it had power to swallow people's souls"[31] can be understood in a new and more literal way today. We might say that the number of people who are dying in landslides or by falling into the holes they have dug is another way of being swallowed by gold. The greed they demonstrate in excavating the soil, breaking rocks, and treating human life so carelessly is another way of losing their souls in the search for gold. Therefore, the traditional belief that gold was sacred and could swallow people's souls if they wore it can be seen as a positive and protective belief, for both the environment and human beings.[32]

The church in the area has not shown a clear position on the issue because of the divergence of opinions among believers. Not only that, but church members are also involved in gold mining and often contribute

---

30. United Nations, "Transforming Our World," para. 2.

31. de Rouville, *Organisation Sociale des Lobi*, 42.

32. There are numerous stories told about people who even died inside the holes because they wanted to get gold by any means possible, even though they were warned about possible landslides. To add to the tragedy, there have also been people who come to mine and end up dying without having any of their relatives' contact information with them. On 20 April 2017, Oho Marguerite Dah, a pharmacist at the Gaoua Regional Hospital, related that a new mining site was opened at Bantara, a village in Kampti in Southwest Burkina Faso. A young man from Goursi (a different area of Burkina Faso) went to work there and was taken to the hospital. He passed away in the course of the week, but did not have emergency contact information on his person. Therefore, the municipal authorities of Gaoua had to bury him.

financially to the church. This disengaged attitude by the church may be seen as a failure, and furthermore is irresponsible in terms of the church's responsibility to protect humanity and all of creation, which God has entrusted to us as stewards. It has allowed the desacralization of gold to swallow up the souls of the churches. God gave dominion to human beings over creation, but dominion should be understood as keeping creation and not the overuse and destruction of nature as can be seen in the area because of gold mining. Even though the church is benefiting from the money that church members who are involved in gold mining bring to the church, it is wrong to be silent over the social injustice that goes on at the mining sites. It is also wrong to be silent over the massive destruction of the earth that is going on and also the defilement of human beings' lives and the land by human beings themselves.

The church should be bold enough to speak up. As Gnanakan suggests, "The Christian community must be willing to identify and condemn social and institutionalized evil, especially when it becomes embedded in systems. It should propose solutions which both seek to reform and (if necessary) replace creation-harming institutions and practices."[33] Thus even though the government, local authorities, and other private institutions such as plan Burkina Faso seem to have proven themselves powerless before the issue of gold mining in the area, it is time for the church to speak up for a change. As Gnanakan further notes, "Growing greed in our consumer-driven society is widening the gap between the exploiter and the exploited and needs to be arrested before it is too late. It is a biblical theology that will help expose and resolve these root problems."[34]

Conclusion

The main goal of the current paper is to create awareness for the new generation to reconsider their culture in relation to creation. In choosing to look at the impact of gold mining on creation in Southwest Burkina Faso, alongside the desacralization of gold, I explored how the underestimation of cultural values can lead to serious environmental consequences which affect the entire creation.

It is thus time for the church to stand up and biblically address the devastation that is taking place against creation in the area. Otherwise, the Christian community will also bear responsibility for contributing to the destruction of creation by keeping quiet, for silence is tantamount to approval.

33. Gnanakan, *God's World*, 206.
34. Gnanakan, *God's World*, 6.

Bibliography

Archives Nationales de Côte d'Ivoire, Abidjan (ANCI-A). Monographie de Bondoukou, Colonie de la Côte d'Ivoire, Haute Côte d'Ivoire Orientale (deuxième cahier, troisième cahier, quatrième cahier), Les Etats de Bouna, par le Lieutenant Chaudron du 1er Régiment de Tirailleurs Senegalais-Commandant la Circonscription de Bouna, 1902.

Assohoto, Barnabe, and Samuel Ngewa. "Genesis." In *Africa Bible Commentary, A One-Volume Commentary Written by 70 African Scholars*, edited by Tokunboh Adeyemo, 9–84. Nairobi: WorldAlive, 2006.

Blasu, Ebenezer Yaw. "Christian Higher Education as Holistic Mission and Moral Transformation: An Assessment of Studying Environmental Science at the Presbyterian University College, Ghana and the Ecclogical Thought of the Sokpoe-Eve for the Development of an African Theocology Curriculum." PhD diss., Akrofi-Christaller Institute, Akuapem-Akropong, 2017.

Bookless, Dave. "To Strive to Safeguard the Integrity of Creation and Renew the Life of the Earth." In *Mission in the 21st Century: Exploring the Five Marks of Global Mission*, edited by Andrew Walls and Cathy Ross, 94–104. London: Darton, Longman and Todd, 2008.

Dah, Ini Dorcas. *Women Do More Work than Men: Birifor Women as Change Agents in the Mission and Expansion of the Church in West Africa - Burkina Faso, Cote d'Ivoire and Ghana*. Akropong-Akuapem, Ghana: Regnum Africa, 2017.

Dah, Sié Eric. 'L'Eglise, Champ de Dieu: Intelligence du Mystère à partir de l'Image du Champ en Milieu Birifor.' Mémoire en Théologie, 6ème Année, Koumi, Juin 1998, Unpublished dissertation.

Daneel, Marthinus L. *African Earthkeepers, Wholistic Interfaith Mission*. Maryknoll, NY: Orbis, 2001.

Delafosse, Maurice. *Les noirs de d'Afrique*. Reprint. Lexington, KY: University of Michigan Library, July 2013.

De Rouville, Cécile. *Organisation Sociale des Lobi, Une Société bilinéaire du Burkina Faso et de Côte d'Ivoire*. Paris: Harmattan, 1987.

DeWitt, Calvin B. "To Strive to Safeguard the Integrity of Creation and Renew the Life of the Earth." In *Mission in the 21st Century, Exploring the Five Marks of Global Mission*, edited by Andrew Walls and Cathy Ross, 84–93. London: Darton, Longman and Todd, 2008.

Fiéloux, Michèle, et al. *Images d'Afrique et sciences sociales, Les pays lobi, Birifor et dagara (Burkina Faso, Cote d'Ivoire et Ghana), Actes du Colloques de Ouagadougou (10-15 décembre 1990)* (Editions Karthala et Orstom, 1993).

Gnanakan, Ken. *God's World: Biblical Insights for a Theology of the Environment*. London: Society for Promoting Christian Knowledge, 1999.

Kalu, Ogbu U. "The Sacred Egg: Worldview, Ecology, and Development in West Africa." In *Indigenous Traditions and Ecology*, edited by John A. Grim, 225–48. Boston: Harvard University Press, 2001.

Kambou-Ferrand, Jeanne-Marie. *Peuples Voltaïques et Conquête Coloniale 1885-1914, Burkina Faso, Préface de Joseph Ki-Zerbo*. Paris: Editions l'Harmattan, 1993.

Labouret, Henri. *Les Tribus du Rameau Lobi*. Paris: Institut d'Ethnologie, 1931.

Lehtonen, Markku, et al. *Le Développement Durable: Signification et Enjeux. Sustainable Development: Meaning and Challenges.* Groupe Caisse des Depots et le Cercle des Economistes, 2002.

Maxime, Compaoré. 'L'école au quotidien en Haute-Volta pendant la période colonial.' In *la Haute-Volta Coloniale: Témoignages, recherches, regards,* edited by Gabriel Massa and Y. Georges Madiéga, 361–77. Paris: Karthala, 1995.

Parrinder, Geoffrey. *African Traditional Religion.* 3rd ed. London: Sheldon, 1974.

Père, Madeleine. *Villages en Transition,* Tome II. Laval Cedex: Edition Siloë, 1988.

Stott, John. *Issues Facing Christians Today: New Perspectives on Social and Moral Dilemmas.* London: HarperCollins, 1990.

United Nations. "Transforming Our World: The 2030 Agenda for Sustainable Development, A/RES/70/1." https://www.diplomacy.edu/transforming-our-world-2030-agenda-sustainable-development.

Whiteman, Darrell L. "Bible Translation and Social and Cultural Development." In *Bible Translation and the Spread of the Church: The Last 200 Years,* edited by Philip C. Stine, 120–41. Leiden: E. J. Brill, 1990.

Part IV

Akan Traditional Perspectives of Land Care

Kofi Agyekum
University of Ghana, Legon

Abstract

This paper discusses the concept of land and land use from the Akan traditional perspective. It considers issues such as land ownership, tracing it to God, Mother Earth, the ancestors, chiefs, families, and individuals. It also highlights land lease and care for land passed on from a group or individual to another. The semantic lexicon on land, agriculture, maxims, idiomatic expressions, and Akan proverbs on land are used to support the Akan concept of land. The paper hinges on cultural and language ideology.

Introduction and Concept of Land

LAND IS ONE OF the most important factors of production, especially in Africa and Asia. It is limited and can hardly be increased, but can highly deteriorate. That is why land care, acquisition, protection, preservation, and documentation are crucial. Land is an indispensable aspect of the Akan culture, and in view of this it is attached to chieftaincy and power. Every powerful chief or king has vast land, normally beyond his geographical location. For instance, the Asante king, Otumfoɔ, has a lot of land in Brong Ahafo, especially the Ahafo portion of the region. The Okyehene of Abuakwa, Osagyefo, has land that extends miles away from the headquarters in Kyebi to places like Kade, Akwatia, and Asamankese. Based on the notion that land cannot be increased except by usurping, buying, or being given as a gift, the Akans do not joke about land because when it is lost, it is lost for good. It is thus not surprising that in pre-colonial times kingdoms waged war on others with the intention of capturing them and taking over their land. This brings the following proverb to light: *ɔhene a ɔnni asaase no na wanko*, "A chief who did not fight has not got land," and again, *wasaase tɛtrɛtɛ na kyerɛ wo tumi ne wo dɔm*. "The extent of your land denotes your power and the people you have."

Language and Cultural Ideology

Ideologies are shared and predictable beliefs and ideas of a people that are noted to be real and implicit in their everyday life situations within a certain period of time.[1] In ideological studies, meaning is socially constructed and to understand people's behavior in social interaction it is better to know their sociocultural, political, and historical backgrounds. Hall stipulates that "ideologies are mental frameworks—the languages, the concepts, categories, imagery of thought, and the systems of representation—which different classes and social groups deploy in order to make sense of, define, figure out and render intelligible the way society works."[2]

A society's ideological concept of any aspect of their life makes them see the item as socially owned, unchangeable, inevitable, indispensable, real, and natural, and they will always justify and rationalize its existence.[3] Culture and society have absolute power over the individual as far as ideology is concerned. In society, ideologies operate as systems of linguistic practices.[4]

In linguistic and cultural ideologies, the beliefs are shared and predictable when it comes to their application and execution.[5] Language and cultural ideologies provide a sociocultural understanding and interpretation of the political, cultural, ecological, economic, legal, and religious processes, as well as people's ways of life that inform the local beliefs about language, culture, and environment. It is upon the multiple functions of the cultural and language ideologies that we are using it as the framework to analyze the concepts and proverbs on land use and care.

---

1. Agyekum, "Sociocultural Concept of *ohia*," 163–71; Agyekum "Role of Language in Conflicts," 18–33; Agyekum, "Language, Gender and Power," 165–83.
2. Hall, "Problem of Ideology," 26.
3. Makus, "Stuart Hall's Theory of Ideology," 500.
4. Makus, "Stuart Hall's Theory of Ideology," 503.
5. Rumsey, "Wording, Meaning and Linguistic Ideology," 346.

The Akan People

The word "Akan" refers to both the people and their language. The Akans are the largest ethnic group in Ghana. According to the 2010 national population census, 47.5 percent of the Ghanaian population is Akan and about 44 percent of the population speaks Akan as nonnative speakers.[6]

The Akans are found predominantly in southern Ghana. Akan is spoken as a native language in six out of the ten regions in Ghana, namely, Ashanti, Brong Ahafo Central, Eastern, Volta, and Western Regions. The Akan-speaking communities in the Volta Region of Ghana are sandwiched by the Ewe communities. The Akan language has 13 dialects, namely, Agona, Akuapem, Akwamu, Akyem, Asante, Assin, Bron, Buem, Fante, Denkyira, Kwawu, Twifo, and Wassaw. Some Bron speakers are found in Cote d'Ivoire. Akan is studied from primary school up to the university level.[7]

The Concept and Ownership of Land among the Akans

In discussing land in Ghana, Peter Sarpong[8] first looked at the religious perspectives of land and saw land as a cognate of Mother Earth. In Akan, the term for Mother Earth is *Asaase Yaa*[9] or *Afia*[10] among Twi and Fante speakers, respectively. It is thus not surprising that the same lexicon *asaase* refers to land. It semantically means that if Mother Earth is a great goddess, then land (*asaase*) is also powerful and a deity. The Akans therefore offer sacrifices to the land, to keep it pure, sacred, revered, and holy, and have taboos that relate to it.

To cultivate the land, especially if the land is a virgin forest, one has to pour out libations to request use of the land. Some flora and fauna, as habitats of the land, are considered spiritual, sacred, and powerful, and must not be troubled. Examples are the *odii* tree and the creeping plant *homakyɛm* (*canthium hispindum*) and the *tweneduro* (*cordia irvingii*), the tree used in carving traditional drums. These plants and trees were considered sacred, and as children of Mother Earth they must be revered. Again, they were scarce and should be protected from extinction. Making sure that scarce things are well protected is a virtue. Another important reason for not

---

6. *2010 Population and Housing Census*, 4.
7. Agyekum, "Sociocultural Concept of *ohia*," 165.
8. Sarpong, *Ghana in Retrospect*.
9. *Yaa* is the Akan name given to a female born on Thursday.
10. *Afia* is an Akan name given to a female born on Friday.

cutting some of these trees was the belief that since they are sacred and powerful, they have medicinal powers and can be used extensively for herbal medicine, and hence must be protected. The medicinal value of plants is still valid enough that when weeding in farms, and even in homes, there are some plants Akans will always leave behind and conserve.

In villages and in farms the Akans made sure that there were some trees left as shady trees. These were left to preserve the land by protecting it from direct sun rays that may be harmful to the plants on the farm. The reserved trees were meant to provide some shade for the plants and for the people in the village. They were also reserved so as to check storms and strong winds that tended to rip the roofs off buildings. On the farms, they were made to serve as windbreakers so that strong winds would not destroy crops like plantain and maize.

Some animals and fish are tabooed by the people since they are considered the children of the land. The bigger animals, like elephants, buffalo, black duiker, and bulldogs were considered *sasaboa*, "spiritual animals," and thus when they were accidentally killed the hunters had to perform spiritual purification rites, or else they would be haunted by their spirits.

Land Ownership, Tenure, and Accountability of Land

The supreme owner of the land is *Onyankopɔn*,[11] "the Almighty God," and the Akans refer to God as *Onyankopɔn a ɔbɔɔ ɔsoro ne Asaase* "God the Almighty who created the Heavens and the Earth." This means that the land belongs to God. This is manifested when there is a land conflict where one could pose the question, *Woreba asaase yi so, wode asaase bae, woabɔ asaase da?* "When you were coming to the earth did you bring any land, have you created land before?"

Before the introduction of Western and Eastern religions, the Akans knew *Onyame*, "God," and what God could do. Sarpong, however, opines that since the Supreme Being and Mother Earth are held to be too remote, the ancestors are the real landowners who continue to take lively interest in the land.[12] The chiefs thus deputize for the ancestors, and in order to sell land, the chiefs and family heads must ask for permission.[13] Authority is graphically seen as:

---

11. At times the name of God is given as *Onyankopɔn Kwame*, where "Kwame" is the name given to a male born on Saturday.

12. Sarpong, *Ghana in Retrospect*, 117.

13. See Nukunya, *Tradition and Change*, 95.

God———————— Ancestors——————
Chiefs————— groups/individuals

There are various types of land titles and stewardship among Ghanaians over customary land. These are allodial titles and freehold titles. There is also temporary stewardship and land care, called *nhwɛsoɔ* stewardship by the Akans, especially among cocoa farmers. There is sharecropping labelled as *abunu* and *abusa* tenancies. The farmer gets half or two-thirds, respectively, based on the quality of the farm, *abunu* (Lit. "break into two") and *abusa* (Lit. "to break into three").

Acquisition of Land and Land Title

Acquisition of land was by purchasing and lease, forceful demand and occupation after a conquest in war, exploration of new-found land by hunters, or a reward to functionaries at the chief's court.[14] The reward aspect of land acquisition is manifested in the following aphorisms about land:

> *ɔsompa de asaase ba.* "Good services bring about land."
>
> *Awarepa ho agyinamdie baako ne asaase.* "One of the sureties of good marriage is land."

Traditionally, the purchasing of land calls for witnesses and rituals to finalize the deal. They include the actual payment, some token for the witnesses, schnapps to be used in pouring libation, and finally *dwahateɛ*, "a testimonial or witnessing system involving the breaking of a twig between the buyer and the seller." The part of the twig which goes to these parties acts as the documentary object that must be tightly kept and be produced as and when necessary.

With Westernization, urbanization, and large-scale farming, the culture of purchasing land for agriculture, industries, business enterprises, educational centers, and housing has rather become the norm. The acquisition through war conquest has been knocked out and hardly do people now give land as a gift.

The Concept of Ɔdehyeɛ, "Royal"

The term *Ɔdehyeɛ* is derived from Ɔ "s/he" / "owner" *de* "to bear/hold/own" and *hyeɛ*, "boundary." The composite refers to the owner of boundaries.

---

14 See Sarpong, *Ghana in Retrospect*, 118.

By semantic extension the owner of the boundaries is the land owner who knows the extent of the land and with whom he/she shares boundaries. In Akan culture, the royal family are said to be the one who settled on the land before all other settlers joined them and hence showed those who joined portions of their land how to occupy and use them. That is why the settlers owe allegiance to the chief and his royal family, who are the *asaase wura*, "custodians/owners of the land." This system is not peculiar to Akans; it is a national phenomenon, if not African or global. The allodial interest is the highest customary land title in Ghana and is held by stools (skins), sub-stools, clans, families, as well as individuals.

There is land care and systems of accountability, even though the king, chief, or *abusuapanin* "family head" is the custodian of the land under his jurisdiction. A violation of this can lead to dethronement or taking his power from him. Each of these custodians is accountable first to his ancestors who previously owned the land. He should therefore follow the traditional rites of tilling the land or transferring the land either through selling, lease, *akyɛdeɛ*, "gifts or gratuities offered out of one's volition" or using it as a surety to ward off some pressing demands in the family.

When land is given as a gift, tradition has to officialize it through the following terms: *aseda*, "appreciation," and *nsatugusoɔ*, "offering of drinks." The appreciation is done by presenting some token of money, sheep, chicken and eggs, or even clothes. The drinks are used to seal the deal, and some are used in the pouring of libations for the ancestors, and also shared among the gathering to indicate that they are witnesses, especially when there are no documents for the transaction. When all the rituals have been performed, the Ashantis use the term, *wɔate asaase no taama ama obi*, "they have handed the land officially to another person."

The chief or family owes it a duty to be transparent and accountable to his subjects and family members. For instance, in the matrilineal system, the *abusuapanin*, "family head" (father), cannot give any family land to his children until it is accepted by his matrilineal kinsmen. He should protect the land spiritually by following the taboos that relate to the Mother Earth and the land. For instance, before cultivating the land, certain sacrifices, including the offering of sheep, chicken, eggs, and mashed plantain or yam *(etɔ)*, are offered.[15] Libations are invoked to call forth the spirits of *Onyankopɔn*, "God," *Asaase Yaa*, "Mother Earth," *Nananom nsamanfoɔ*, "ancestors," the river deities, and all other supernatural beings in the ecological environment to protect the land and the people who are going to cultivate it.

---

15. See Sarpong, *Ghana in Retrospect*, 116.

The custodian makes sure that no criminal activities occur on the land, e.g., people having sexual intercourse, committing suicide, cursing someone, or fighting in the bush, etc. People should avoid going to farm on sacred days like *Fofie, Adae, Fɔdwoɔ*, etc. Again, people should avoid cursing people while on their farms. The violation of all these taboos will result of lower agricultural yield. This has brought the euphemism, *ɔdɔ Benada* to stand for an "impotent man." Elsewhere I averred that 'In some Akan communities, people do not go to farm on Tuesdays, for Tuesday is a sacred day and people are barred from working on their farms . . . People who farm on Tuesday do not have any proper harvest."[16]

When a royal family occupies a new land, hunters play a tremendous role. They functioned as land surveyors who knew the length and breadth of the land. They had mastery of the topography and ecology of the land, including water bodies, rivers and streams, mountains, hills and valleys, categories of soils (alluvial and sedimentary,) rocks, and stones. As explorers, they knew the streams and rivers that had constant water, even during dry seasons, and which rivers flood during rainy seasons. Information about all these ecological changes was vital for the royal family. It is for these reasons that in *abɔfoɔdwom*, "hunters' song," the hunters use innuendoes to pride themselves and comment about their neglected usefulness in the community.[17]

Land Care and Sanitation

The Akans traditionally put religious, social, and administrative measures into care for the land both at home and in the bush. Politically, the *abusua panin*, and sometimes the chief, determine which lands have been left to fallow for some time that when tilled there will be an abundant harvest. He also saw to it that the land was fairly distributed among the family members. In cases where lands had not been left to fallow for a long time, the family members were advised to use *porɔka*, "a traditional manure," i.e., where after weeding and felling the trees, the land was not burnt, but the bushes were allowed to get rotten and the food plants were planted through them.

On the contrary, if the land has to be burnt, Akans are very careful that the bush fire not extend beyond that farm land to burn other people's farms. They use the expression, *pa afuo no ano*, "where one weeds to the bare ground at the edges of the farm," to keep the fire from escalating. The farmers make sure that they ask colleague farmers to accompany them to

16. Agyekum, *Akan Verbal Taboos*, 181.
17. Nketia, *Abɔfodwom*.

burn the farm, hence ɔbaako nhye afuo, "one person does not burn a farm." The Akans therefore have it that communal help in land care is the ideal situation, hence the term nnɔboa, "cooperative farming system," among rural Akans. Farmers are able to arrange so that they can go and weed in farmer A's farm this week, and shift to B's farm the following week. Cocoa farmers also plan in the same manner during harvesting so that a single farmer is not pressurized by a lot of work.

Sanitation

In terms of sanitation, the traditional Akans are very cautious about protecting their villages, streets, homes, water bodies, and forests. In view of this, Akans have superstitions that bar people from defecating or doing laundry within a certain distance from their streams and rivers. They do not want to pollute the water bodies, which are their only sources of water for drinking and other domestic purposes. Again, people are not allowed to farm closer to water bodies, or cut trees that were near their water bodies, for fear of having direct sunshine in the streams and thereby drying them and denying the people of their source of water. The big rivers are considered deities, hence the term *asubosom*, "water deity"; they are sacred, revered, and have an established sacerdotal system.

In the villages, sanitation was so crucial and part of the responsibilities of children, adults, and the entire community. Chiefs made sure their villages were always neat. People were obliged to sweep their houses and weed around their surroundings, as well as the closer environments. Every village had a refuse dump and occasionally there was communal labor where all the women would go and clear the place and burn the garbage. The children in the village were grouped into boys and girls and they swept and weeded the streets and the public toilets. The men made sure that they weeded the path to the farms and the streams and also dredged the streams to get clean water.

All these were communal responsibilities to provide serene environments to prevent certain diseases. Anybody who failed to attend to such communal labor was fined and if s/he refused payment, he/she was dragged to the chief's house for a tougher punishment and severe fines. All the above indicate the civic responsibilities of all the people to protect the land, adhere to sanitation, and live healthily and comfortably on the land.

Proverbs and Aphorism on Land and Land Care

Proverbs are general truths, principles, and ways of life that are based on people's past experiences, mindsets, philosophies, perceptions, ideologies, sociocultural concepts, and world views.[18] Oral literature scholars have researched extensively into proverbs. Agyekum opines that "proverbs are interpretations of traditional wisdom based on the experiences and sociocultural life of our elders."[19]

Agyekum further posits that "In the Akan context, there are adages such as *ɛbɛ ne ɔkasa mu abohemmaa*, "the proverb is the most precious stone of speech," *Etwa asɛm tia*, "it curtails matters," and *ɛnka asɛm ho a, ɛnwie dɛ yɛ*, "without it, a speech does not acquire its sweetness."[20] The indigenous Akan perceives the proverb as an indispensable and aesthetic device that is vital in speech, and as the salt of a language, without which the real taste of the language dish is not felt.

Ssetuba avers that in Africa, "The proverb is regarded as a noble genre of African oral tradition that enjoys the prestige of a custodian of a people's wisdom and philosophy of life."[21] Proverbs are the analytic tools of thought, when thought is lost; it is proverbs that are used to search for it.[22] Finnegan states that "In many African cultures, a feeling for language, for imagery, and for the expression of abstract ideas through compressed and allusive phraseology comes out particularly clearly in proverbs."[23] The proverbs on Akan land reflect Akan philosophies, religious beliefs, and experiences on Mother Earth and the land. Let us look at the following proverbs on land. The proverbs are categorized under ownership and power, resilience of the land, and hard work on the land.

A) Proverbs on Land Ownership and Power

(1) *ɔhene a ɔnni asaase no na wanko*

"A chief who has no land did not go for wars."

(2) *Wasaase tɛtrɛtɛ na ɛkyerɛ wo tumi.*

18 See Agyekum, "Akan Proverbs and Aphorisms," 11.

19. Agyekum, "An Akan Oral Artist," 9. See also Agyekum, "Akan Proverbs and Aphorisms"; Finnegan, *Oral Literature in Africa*; Hussein, "Social and Ethno-cultural Construction," 61; Oluwole, "Culture, Gender, and Development Theories," 95–121; Okpewho, *African Oral Literature*; Ssetuba, "Hold of Patriarchy"; Yankah, "Proverbs," 325–46.

20. Agyekum, "An Akan Oral Artist," 10.

21. Ssetuba, "Hold of Patriarchy," 1.

22. See Oluwole, "Culture, Gender, and Development Theories," 100.

23. Finnegan, *Oral Literature in Africa*, 380.

"The extent of your land depicts your power."

(3) *Penten asaase ɛnna hɔ kwa.*

"Penten land is not lying there as a fallow land."

(4) *Nkokɔhwedeɛ mmienu nkɔko obi afuo ase.*

"Two partridges do not quarrel in somebody's farm."

The first two examples above relate land ownership to power and war conquest. In those days, kings and chiefs who were able to fight wars and conquer were very powerful and also had human, financial, and material resources.

Examples three and four indicate that there is no land without an owner. The fact that you do not see people tilling the land does not mean that it is ownerless and free to be used by anybody. Example three indicates that no land lies fallow; it has a use. Probably the opportune time has not arrived yet. Example four comments on people who fight over lands that are not theirs and thus advises on unnecessary land disputes since the real owner is there.

B) Resilience of the Land

(1) *Asaase nyɛ kɛtɛ na yɛabobɔ.*

"The earth is not a land to be folded."

The above indicates how solid and resilient land is and thus must not be trampled upon. It is not like a mat where you can decide to spread and fold anytime you want.

(2) *Asaase tokuro obiara bɛkɔ mu.*

"Everybody will enter into the earth's hole."

The above proverb sees land as the final journey of man, and the fact is that everybody will die and be buried. It indicates the magnanimity of the land/earth, which is receptive to everybody, rich or poor, black or white, tall or short.

C) Proverbs on Hard Work on the Land

The proverbs below allude to hard work and full utilization of a piece of land

(1) *Yɛntena mfofoo ho mma no nnane yɛn kwaeɛ.*

"We do not sit behind the bush for it to turn into forest."

(2) *Deɛ ɔkeka adwooguo pii no ntwa mfudeɛ biara.*

"He who clears land without cultivating does not harvest any yield."

(3) *Wodɔ afuo na Onyame anhunu mu a, afuo no nyɛ yie.*

"If you make a farm and God does not see it, the farm will not yield products."

The three proverbs above comment on the proper care and utilization of the land and eschew laziness and lackadaisical attitudes toward land care and use. The first proverb above states that if one starts any project on land, the person needs to focus and forge ahead. Without focusing, a simple and manageable work like weeding a small bush will turn into a difficult task like weeding a forest with lots of trees. The second example above highlights on focusing when it comes to land care and use. Some farmers will engage themselves with pockets of farms and in the end will not have one complete farm that is highly productive. It is so, too, with housing projects, where you see various uncompleted housing projects sprawled around, none of which is habitable. The third example above hammers on the need for hard work, indicating that if you are making a farm it should be open enough to get rains and sunshine, or else there will be decreased productivity.

Conclusion

In this paper, we have looked at the indispensability of land, looking at it from the standpoint of Akan language and cultural ideology. We found out that the same term *asaase* for Mother Earth is used for land. This means that like Mother Earth, which is considered a major deity, the land is also a deity from the Akan religious perspective. The religious philosophy is manifest when we talk about land ownership, where we see that the land in Akan belongs to the ancestors and that the kings, chiefs, and family heads are only custodians of the land, hence there is land accountability. The paper also discussed the taboos and sanitation associated with land use and care.

The paper has discussed how land was acquired in the Akan traditional setup and we posited that it was through war conquest, gifts from land owners, settlement patterns, and purchasing or lease for some period of time. With respect to land exploration, we have found out that hunters played very significant roles in finding new lands, giving proper information about the land, its fertility and general knowledge about the ecology.

Finally, we looked at the Akan concept of land through the lenses of Akan proverbs, which are based on philosophical truths and the experiences of the Akan over centuries of land use. We discussed proverbs that associated land with the power of the king, hard work on the land, focus on land use, and the resilience of the land.

Bibliography

2010 Population and Housing Census. Accra: Statistical Service of Ghana, 2012.

Agyekum, Kofi. "Akan Proverbs and Aphorisms about Marriage." *Research Review* NS, 27.2 (2012) 1–24.

———. *Akan Verbal Taboos in the Context of the Ethnography of Communication*. Accra, Ghana: Ghana Universities Press, 2010.

———. "An Akan Oral Artist: The Use of Proverbs in the Lyrics of Kwabena Konadu." *Research Review* 21.1 (2005) 1–17.

———. "Language, Gender and Power." *Issues in Intercultural Communication* 3.2 (2010) 165–83.

———. "The Role of Language in Conflicts in Ghana." *International Journal of Political Discourse Analysis* 4.1 (2011) 18–33.

———. "The Sociocultural Concept of O*hia*, 'Poverty' in Akan: Konadu's Song *ɛnneɛ mekɔ na maba*." *South African Journal of African Languages* 36.2 (2016) 163–71

Finnegan, Ruth. *Oral Literature in Africa*. 2nd ed. Cambridge: Open Book, 2012.

Hall, Stuart. "The Problem of Ideology: Marxism Without Guarantees." In *Stuart Hall: Critical Dialogues in Cultural Studies*, edited by David Morley and Kuan-Hsing Chen, 24–45. London: Taylor & Francis, 2005.

Hussein, Jeylan W. "The Social and Ethno-cultural Construction of Masculinity and Femininity in African Proverbs." *African Study Monographs* 26.2 (2005) 59–87.

Kroskrity, Paul V. "Language Ideologies." In *A Companion to Linguistic Anthropology*, edited by Alessandro Duranti, 496–517. Malden, MA: Blackwell, 2006.

Makus, A. Anne. "Stuart Hall's Theory of Ideology: A Frame for Rhetorical Criticism." *Western Journal of Speech Communication* 54 (1990) 495–514.

Nketia, Joseph H. K. *Abɔfodwom*. Tema: Ghana Publishing, 1973.

Nukunuya, George K. *Tradition and Change: An Introduction to Sociology*. Accra, Ghana: Ghana Universities Press, 2003.

Oluwole, Sunday B. "Culture, Gender, and Development Theories in Africa." *Africa Development* 22.1 (1997): 95–121.

Okpewho, Isidore. *African Oral Literature: Backgrounds, Character, and Community*. Bloomington: Indiana University Press, 1992.

Rumsey, Alan. "Wording, Meaning and Linguistic Ideology." *American Anthropologist* 92.2 (1990) 346–61.

Sarpong, Peter. *Ghana in Retrospect. Some Aspects of Ghanaian Culture*. Tema: Ghana Publishing, 1974.

Silverstein, Merril. "The Uses and Utility of Ideology. A Commentary." In *Language Ideologies: Practice and Theory*, edited by Bambi B. Schieffelin et al., 122–45. New York: Oxford University Press, 1998.

Ssetuba, Isaac. "The Hold of Patriarchy: An Appraisal of the Ganda Proverb in the Light of Modern Gender Relations." A Paper for Cairo Gender Symposium, April 7–10, 2002.

Yankah, Kwesi. "Proverbs: The Aesthetics of Traditional Communication." *Research in African Literature* 20.3 (1989): 325–46.

The Krobo Religious View of Land Use and Care

Robert Mate Wayo Opata

Abstract

Land use and care is regarded as sacred in the religious sense of the Krobo. The religious view of Africans and/or the Krobo is seen both personally and corporately through belief in Mawu, the Supreme Being, in the observance of land rituals and harvest festivals. The relations between God, humankind, and care for the land uphold the existence of life and well-being of the people. Life on the land will become a burden and disastrous when the relationship between God, humankind, and the land are not kept sacred by its users. Therefore, this paper highlights the Krobo concepts in which land is religious and sacred. The paper seeks to argue that in tracing the features that influence land use one must look at the religious sense of its use and care.

Introduction

AFRICAN TRADITIONAL RELIGION SETS patterns of behavior for the community as a whole, thus influencing individual behavior and society.[1] In recent times, the sacred idea of land use and care is not familiar to the industrial and commercial-minded people in Ghana. This paper will show the Krobo religious views, use, and care for the land, with a particular focus on the sacred use. The religion of the Krobo teaches that land is a sacred trust of

1. Quarcoopome, *West African Traditional Religion*, 163.

God to the ancestors; selling of ancestral property was ritually forbidden in the olden days.[2] Special rituals took the place of indentures of our time when land was transferred to another. These ancient rituals and practices of land regulations and agreements are partly in use today. The Krobo religious values and beliefs have a great impact on the way the people think and act in relation to their environment which brings either harm or good fortune. The main features of the traditional religious sense of the Krobo, to a large extent, centers on their relationship with *Mawu* (God), the deities, and land cultivation.[3] A good relationship with the land leads to fruitfulness and the ability to germinate seeds. The fertility of the land is important to the Krobos, who are farmers, as they grow their produce. For farmers, their well-being is largely determined by agricultural harvests and not necessarily by their commercial activity. The religious traditions of the Krobo are thus great treasures of ancient wisdom.

This paper basically focuses on Hugo Huber's *The Krobo*.[4] Although others have tried to record the proceedings of the Krobo social life using accurate first-hand data, we acknowledge that no comprehensive and detailed inquiries have been made into the social changes taking place regarding the sacred and nonsacred use of land in recent times.[5] This paper will highlight the religious understanding of the Krobo and show the perspectives on how the use of and care for the land can be done through the religious system.

Religion, Land, and Cultivation

John S. Mbiti, in his book *Concepts of God in Africa*, demonstrated from various perspectives that God is conceived in African religion to be the Supreme Being who created the whole universe.[6] He shows how the African belief lies in a deep faith in God working as a spirit who oversees the earth. God, who is in direct relationship with the land, river, and plants, has his abode on sacred mountains and sends forth rains to water and nourish the land.[7] For Peter Sarpong, one cannot talk about the land and its first owner without pointing out its main religious features.[8] The Krobo, as indeed do

---

2. Huber, *Krobo, Traditional Social and Religious Life*, 40–41.
3. Conteh, *Essays in African Religion*, 123–25.
4. Huber, *Krobo, Traditional Social and Religious Life*, 7.
5. Huber, *Krobo, Traditional Social and Religious Life*, 7.
6. Mbiti, *Concepts of God in Africa*.
7. Mbiti, *Concepts of God in Africa*, 145.
8. Sarpong, *Ghana in Retrospect*, 117.

all Africans, believe in the existence of one Supreme Deity who they call *Mawu*—the Creator of all things; the land belongs to him.[9]

Huber discussed the traditional, social, and religious life of the Krobo people in which he "attempts to give an analysis of the traditional aspects of Krobo life."[10] Particularly, it provides a source for interpreting the traditional, social, and religious rules, and ethics of the people. Some of the Krobos descended from *Klowem*, "the mountain top," to the farm lands on the plains, and developed that into a second home. The people living on the mountaintop called the villages or farms on the plains *Dɔm*, "in the valley below." The farm villages on the plains at the foot of Akwapim Ridge were clearly different from their former villages on the mountaintop which they called home (*Klowem*). Some of the Manya and Yilo Krobos have their towns on the mountaintop where they come together as one group on few occasions in a year to meet over issues like defense.[11]

> An important yearly function was '*Atɛ zu wo yo he*' (an operation against erosion). All groups of the tribe went down to the foot of the hill, weeded right round the mountain and scooped up the rich soil, which had been washed down into the valley, to fill up the gorges and ravines on the mountain side. This was a way of saving the mountain settlement from the harmattan wild fires, and checking erosion.[12]

When the Krobos were descending from *Klowem* to the settlement in the plain villages, they carried their religion and social organization with them from the mountaintop to their new settlements or homes in the plains. The system they had on top of the mountain was by priesthood-led communities.[13] The principal deities of the Manya and Yilo Krobos found their sacred places of worship amongst the people. In the course of time, the Krobos' new farm villages and plantations spread all over the fertile lands on the plains and beyond. The villages far beyond the plains are called *yonɔ* or *ngmɔ si*, meaning "villages" or "farm lands." The fundamental occupation of the Krobo was essentially farming. The people cultivated corn, millet, other cereals, and various types of yam for food with the surplus going to the market.[14]

9. Sarpong, *Ghana in Retrospect*, 9–13.
10. Huber, *Krobo, Traditional Social and Religious Life*, 7.
11. Odonkor, *Rise of the Krobos*, 3–4.
12. Odonkor, *Rise of the Krobos*, 52.
13. Odonkor, *Rise of the Krobos*, 52.
14. Abedi-Boafo, *Dangme Nyaii*, 144.

Zugba Zu, "Mother Earth" as it is commonly called, is the name of the earth or the land among the Krobo. When referring to ownership of the land, the term *Nyingmo Zu* meaning "God's land or earth," is used.[15] They believe that *Mawu*, the name of God, is the Creator of the earth and whose wife is the mother earth—*Zugba Zu*. Among the Ugandans also, the earth is the mother of all.[16] The religious belief in *Mawu* as the owner of the land is in no doubt real among the Krobo people. The Krobos believe that the earth is the wife of God, hence the maxim *Mawu kɛ eyo Zugba Zu* "God and his wife, Mother Earth." God and Mother Earth are invoked in most cases, especially in libation prayers, as the first and second owners of the land, respectfully, before the gods, and then the ancestors are invoked to bless the land to feed human beings and the creatures who dwell in it.[17] *Mawu*, the primary owner of the land, feeds and nourishes the inhabitants of the land. Significantly, God and his wife *Zugba Zu* have no priests and shrines among the Krobos.[18]

Fertility and crop increases are by the power of *Mawu*. He has commanded humankind to increase and to spread on the surface of the earth.[19] Traditionally they accord a relative high loyalty to *Mawu* because of "their fear and respect for the sanctions of their ancestors and their deities" in respect to their lands.[20]

The ceremonial transfer of land involves a covenant act by invoking of *Mawu*—the Supreme Being, the gods and the ancestors as witnesses to the agreement between the one transferring and the one receiving it. The two parties will then plant a stone or a tree on the boundary and offer an animal sacrifice there on the site to seal a permanent deal between the two parties.[21] This boundary made of stone or trees should never be removed by anyone. One may be permitted to remove them as long as all the parties can meet and work together. There are serious spiritual consequences that follow illegal removal of boundary lines or stones, especially when the animal is sacrificed on it.[22]

Likewise, traditionally, the sale of ancestors' land and property, or the cutting of palm plantations, requires permission from the ancestor who first

---

15. Quarcoopome, *West African Traditional Religion*, 85.
16. Mbiti, *Introduction to African Religion*, 210.
17. Huber, *Krobo, Traditional Social and Religious Life*, 42.
18. Sarpong, *Ghana in Retrospect*, 16.
19. Mbiti, *Concept of God in Africa*, 60.
20. Huber, *Krobo, Traditional Social and Religious Life*, 294.
21. Huber *Krobo, Traditional Social and Religious Life*, 42–46.
22. Huber, *Krobo, Traditional Social and Religious Life*, 38.

received the land or property or planted the trees. Permission is needed before selling the land or property or cutting down the trees to avoid grave misery and affliction on the family.[23] In a mutual agreement, the head of the family, in the midst of the assembled family members, invokes the name of the ancestor who first acquired it in libation prayer, thus:

> My father or grandfather
> The palm trees which you have planted
> Are too close to each other
> And the young one wants to cut down the trees.
> This is the wine collected for you.
> They have brought it home
> To have it poured out for you in the house.
> Therefore, may nothing harm them.
> May the selling of the land
> Or cutting down of the palm trees
> Bring blessings to us.

In this instance, a full explanation of the reason for the sale of land or cutting down of trees in libation prayers is necessary before one can resort to it. That is to say, the Krobo believe that *Mawu*, the Supreme Being, who is the primary owner of the land, and his wife *Zugba Zu* and the ancestors are to be informed for authorization about the deal in front of all the parties involved.

The sacred agricultural rites are basically related to the natural care for the land. There are ritual laws and prohibitions to be followed like in the case of no farming work once a week on Thursdays among the Krobo. This is observed in direct link with the land, its fertility, and the care for it. Like J. O. Y. Mante avers, the Sabbath rest in the economy of God amongst the Jews is an ecological eschatology like in African traditional religion.[24] The implication of rest on the seventh day for the land does not lie merely in principles of soil chemistry. It lies in the revelation that the holy day is a Sabbath rest both for the Lord and the land. The Jews observed this sacred day of God as religious people for the reason that they are not the sole owners of the land, but they hold the property of land in their possessions in trust under God. They are occupants of the land which God has given them to use and care for it (Lev 25:23–24).

---

23. Huber *Krobo, Traditional Social and Religious Life*, 42–46.
24. Mante, *Africa Theological and Philosophical Roots*.

The religious practice of rest or a prohibition day from farm work once a week is also observed among the Akans. This means the holy days imply a total number of fifty-two days in a year which will allow God's creation of land and vegetation rest from all human activity. The fifty-two days is almost two months of rest, which traditional religions have reserved for the reconstruction of land and vegetation to take place. These holidays help the set of relationships that exist between organisms and their environment. This perhaps describes why the Krobo did not consider farming and care for the land merely as a money-making activity, but as a religious activity that contributed to worship, human ecology ,and the well-being of the environment.

The motherhood of the land appears to be related to the ability to conceive and nurture seeds planted in the soil by farmers. The earth has the power to cause the seeds to incubate and produce food for the sower with which to feed the dwellers on the land: human beings, birds of the air, and animals that walk on the surface of the land. Since agricultural reproduction is likened to divine activity, the religious rituals are necessary. The Krobo religious view of agriculture in relation to the land can be likened to the human reproduction process. Human beings are born from the womb of our mother, and on the other hand we are buried in the womb of the earth.[25]

Planting Rituals and Harvesting Festivals

A ritual is a traditional form of carrying out a religious rite. It is a communication of religious meaning through words and action. At the completion of a ritual act, the people mostly feel as though they have influenced or taken control over the spiritual beings and natural spirits which rule over the land, river, mountains, and trees, among others. The Krobos observe some religious rituals before farming; it is the custom of the people to make sacrifices to the earth before it is cultivated.[26] The Krobos' basic religious beliefs can be found in two farming rituals, which are pre-planting rituals (*koda kpami*) and post-harvesting festival (*ngma yemi*). These sacred religious rituals are important to the use and care for the earth, soil, crops, and the seasons.[27] The first is the ritual of sowing, called *koda kpami,* meaning "cursing or expelling hunger." It begins with a cry of alarm by the priest, referred to as *bubuubui,* with simultaneous responses from all, and on a Monday. The chief priest prays for rainfall and an abundant season of harvest. The preparation

25. Mbiti, *Introduction to African Religion,* 208.
26. Huber, *Krobo, Traditional Social and Religious Life,* 41.
27. Mbiti, *Introduction to African Religion,* 134–35.

for the sowing of seeds was a national affair and every farming season began with a definite religious ceremony led by the chief priest, without which no one had the right to sow seeds:[28]

> This was the ceremony of 'koda,' probably a homage paid to the god of fertility. Between the performance of this ceremony and the harvest season, the Okumo had no right to eat the meat of any bird—wild or domestic. This was strictly observed, because the bird might have stolen a few grains of the unharvested corn. The first-fruit of the farm went to the Okumo, Asaa, Ajime, Atsikpe, the Klowɛki priests. After that the people were free to use the corn, (or whatever crop it might be).[29]

After the sacred pre-planting ritual on Monday, all planting of seeds begins from Tuesday to Saturday until the end of *koda* cry. These sacred religious practices are attached to *Zugba Zu*—mother earth or the land, food crops, farming systems, and land tenure organization.[30] The observation of this ritual influences the coming of rains to renew the activities of the people, vegetation, and animal life. This pre-seasonal religious ritual is an act of renewing, sanctifying, and reviving life for both human beings and other creatures on the land.[31] Through this unique traditional belief and sacred ritual, humanity shows how dependent we are on the natural environment. The Krobos do not use the land at their own will.

The second ritual (post-harvest period) is called *Ngma yemi*, meaning "feasting on millet." This is an annual agricultural festival of the Krobos to represent the beginning of the abundant harvest, where first fruits were offered to *Mawu* and the deities:

> The time of planting and harvesting depends largely on the seasonal periods of the year. There is the dry harmattan season (*hlabata*) lasting from about the end of November to March; it is the time when the fields lie fallow, when the forest is cleared and toward the end of the first crops are planted. In March or April the first rainy season (*gbiɛ*) starts, lasting up to about June; this is the period of some further planting and of weeding. As the last year's food supply is nearly finished by this time, it frequently imposes restrictions on the daily diet. The third season,

---

28. Odonkor, *Rise of the Krobos*, 53.
29. Odonkor, *Rise of the Krobos*, 55.
30. Huber, *Krobo, Traditional Social and Religious Life*, 41.
31. Mbiti, *Introduction to African Religion*, 135.

called *Mau-lɛ* (God gives food) falls between July and September; people feel happy as the first crops are harvested.[32]

This *Ngma yemi* agricultural festival includes the presentation of the harvest to *Mawu* and the deities. It is the ritual of thanksgiving to offer food for *Mawu* and spirits (deities) to eat.[33] Hence, most of the cultivated crops and the foods play an important ritual role in the Krobo agricultural economy. By implication, the religious ceremonies reveal the Kroboss belief that there is a causal relation between human guilt and the sterility of the farm lands, use, and care.[34]

After the annual ritual rites have been performed and the minds of the people cleared from all evil intentions, they have nothing more to do but to look forward to the blessing of God and the gods. What the people expect in the coming season is blessings on the land, fertility of humans and their animals, and good health and riches in their commercial activities. It is clear that the relationship between the land, farming, and religious beliefs are linked together. This sense of relationship actually makes God's creation complete as people use and care for the land.

Mother Earth and Pottery

The Krobos are good in pottery. The center of pottery among the Krobo is Okwenya. This pottery village is located between Somanya and Akuse in the Eastern Region of Ghana. The ancient pottery skill of the women can still be seen today near the Krobo Mountain. *Zu Muɔ*, is the name the Krobo called a clay pit or the spirit that controls the clay which Krobo women use for pottery.[35] Traditionally, they believed that the money used by Krobo farmers in purchasing virgin forest lands for the cultivation of cocoa in Ghana was realized from the sale of pottery.

Krobos have ritual, household, and magical pottery. There is great demand for these three categories of earthenware. The process of making the earthenware looks simple, but the production comes with various rituals. Household pottery is used to contain all kinds of items, food, drinks, and strong medicine. It is a common believed among the Krobos that the potter's wheel upon which the clay is moulded for ritual or magical pottery is

32. Huber, *Krobo, Traditional Social and Religious Life*, 50.
33. Asamoah-Gyadu, *Spirit and Spirits*, 440–43.
34. Huber, *Krobo, Traditional Social and Religious Life*, 253.
35. Abedi-Boafo, *Dangme Nyaii*, 144.

endowed with special powers.³⁶ Thus potters behind the potter's wheel perform certain rituals necessary to transmit power to the clay mould during the production process of these earthenwares, especially for sacred or spiritual purposes. For instance, it is believed that a type of magical pottery called *zene* earthenware is used for black magic. One can bury this magic pottery in another's compound or yard and pronounce a curse on that person and the curse will result in his/her sickness or death. This is possible based on what the potter does. There is also the general rule that for spirits to accept offering in their sanctuary, it must be offered in primitive wares; in old traditional equipment with some techniques, customary rules, and language.³⁷

The Relationship between God, Humankind, and the Land

In his *Africa Theological and Philosophical Roots of Our Ecological Crisis*, J. O. Y. Mante attempts to make a case for ecological atmosphere to be taken seriously as African theologians. Mante points out several burning issues confronting the African continent and attempts to show the theological and philosophical roots of our ecological crisis as a quest for an indigenous theology.³⁸ Mante argues that:

> To have an ecological orientation, then, is to think in such a way that whenever one thinks about humans, one thinks of them instantly within their natural habitat or environment—humans-in-relation, not only with other humans, but also (and at the same time) with the nonhuman environment. Any moment we think of a man or woman, we are to think of them in relation to both their human and nonhuman environment.³⁹

He goes further to state that the nonreligious and nonecological form of living, and that which does not take the rest of the natural environment into consideration, does not just destroy God's creation, but gradually eliminates human existence from the solid surface of the earth.⁴⁰ These perhaps underscore the prevalence of human and animal diseases in the environment and its impact on life expectancy, human capital formation, labor participation, and economic growth. God, in his infinite wisdom,

---

36. Huber, *Krobo, Traditional Social and Religious Life*, 66.
37. Huber, *Krobo, Traditional Social and Religious Life*, 65.
38. Mante, *Africa Theological and Philosophical Roots*, 43.
39 Mante, *Africa Theological and Philosophical Roots*, 12.
40 Mante, *Africa Theological and Philosophical Roots*, 12.

created the land to produce food and nourishment to all living creatures. However, natural disasters like drought and floods caused by human activities worsen already difficult developmental challenges of our environment in Africa. In a sense, by disregarding the management of land and its natural mineral resources, it could be said that human beings are the enemy of the environment. Instead of using the land's rich natural resources and endowment for development, human beings have allowed it to become an instrument of self-destruction.

In African traditional religion, it is held that God created the heavens and the earth and all that is within them. God is so acknowledged as a giver of life during libation prayers, and when sacrifices are offered to the minor deities. God is thought of as the life-giver, the producer of rain and sunshine. The human relationship with God is on the same model of a human relationship.[41] Human beings depend on all the things God has created for their use. Hence, the production of food for human survival must be in relationship with the land and with God who gives good harvest. The Krobo farmers understand that seeds that produce good food to sustain human life come from God. Such a view creates a tripartite religious relationship between God, man, and the land, and further creates, promotes, and cements fruitful union. For the Krobo, plants protected by God are never hurt by the wind.[42]

Breaking such a tripartite relationship means famine, hardship, and disaster for humans. It is said that the penalty for a breach of land prohibitions could be death because the bad behavior of an individual could bring severe cost to the whole community.[43] A Nigerian proverb puts it this way: "whenever a person breaks a stick in the forest, let him consider what it would feel like, if it were himself that was thus broken."[44]

Through human activities against the land, the heavens will not produce its rain, the land will not yield its fruits, and human life will not improve. Poverty, drought, and famine are some consequences believed to be God's punishment to those who disobey God's plans and purpose for the land. As individuals and the community disregard the sacredness of God's intentions toward the land and care, human beings are nothing on the land. A Nigerian proverb says: "All that we do on earth, we shall account for knelling in heaven."[45]

41. Sarpong, *Ghana in Retrospect*, 116.
42. Mbiti, *Introduction to African Religion*, 208.
43. Sarpong, *Ghana in Retrospect*, 116.
44. Mbiti, *Introduction to African Religion*, 208.
45. Mbiti, *Introduction to African Religion*, 208.

Conclusion

African traditional religion essentially centers on their religious view of God, the ancestors, humankind, the land, and its use for farming. Among the Krobos, *Mawu*, the first owner of the land nourishes the inhabitants of the land. The traditional belief of the Krobos is that *Zugba Zu*, the Mother Earth, is the wife of *Mawu*, the Supreme Being. Land is conceived of as a sacred trust of the ancestors who, when they were alive on earth, kept it for themselves and their children. The agricultural rituals of the Krobos before and after cultivation show human dependency in relation to the land. It is believed that *Mawu* and the spirits that inhabit the land, soil, and trees promote fruitfulness and will avert all harm before cultivation. The discussion of the Krobo religious view and the care for the land has revealed that the relationship between God, humankind, and the land promotes and improves the well-being of humanity on the earth. The Krobos' religiosity toward land use actually makes God's creation complete through farming and care for the land. However, there are severe consequences for exploiting the land of God and the ancestors which has been given to humankind as trustee and all will reap what they sow on the land.

Bibliography

Abedi-Boafo, John. *Dangme Nyaii*. Accra, Ghana: Bureau of Ghana Languages, 1978.
Conteh, Prince S. *Essays in African Religion and Christianity*. Accra, Ghana: Cynergy Media 2014.
Gyekye, Kwame. *Philosophy, Culture and Vision: African Perspective*. Accra, Ghana: Sub-Saharan, 2013.
Huber, Hugo. *The Krobo, Traditional Social and Religious Life of a West African People*. Vol. 16. Bonn: The Anthropos Institute, 1973.
Mbiti, John S. *African Religions and Philosophy*. New York: Doubleday, 1993.
———. *Concepts of God in Africa*. London: SPCK, 1982.
———. *Introduction to African Religion, Second Edition*. Oxford: Heinemann Educational, 1991.
Odjidja, Edward M. L. *Mustard Seed, The Growth of the Church in Kroboland*. Accra, Ghana: Waterville, 1973.
Odonkor, S. S. *The Rise of the Krobos*. Tema: Ghana Publishing, 1971.
Opoku, Kofi Asare. *West African Traditional Religion*. Lagos, Nigeria: FEP International, 1978.
Quarcoopome, T. N. O., *West African Traditional Religion*. Ibadan, Nigeria: African University Press, 1987.
Sarpong, Peter. *Ghana in Retrospect, Some Aspects of Ghanaian Culture*. Accra, Ghana: Ghana Publishing, 1974.

Reflections

Reflections: A Ghanaian Christian View of Land Care

ALLISON M. HOWELL

Akrofi-Christaller Institute of Theology,
Mission & Culture, Akropong

Introduction

IT IS DIFFICULT TO clearly identify a single Ghanaian Christian view of land care. The authors of this volume evidently hold their own definitive views on the importance of caring for the land. They also illustrate, however, that some Ghanaian Christians, knowingly or ignorantly, show little respect for the land. The citizens contribute thus to significant prevailing problems related to land care. This attitude also reveals their different views of the land care.

In these reflections, I attempt to distil from the collection of papers what we may consider as a composite Ghanaian Christian view of land care. I also incorporate the perspectives of three other Ghanaian Christians, to the purpose of underscoring the significance of developing theological instruction at all levels within the Christian church. This process would span not only the highest levels of Christian training, but also the often highly neglected area of practical teaching and the training of children and adolescents within the church. To sharpen the discussion in these reflections, I have identified four key areas that provide a possible framework for a Christian view of land care.

1. Understanding key terms

For a clear Ghanaian Christian view of land care, and to avoid confusion, it is important to understand some of the key terms in use. English terms such as "earth," "land," "environment," "ecology," and "ecotheology" are used frequently in this volume. Even so, how do Ghanaians explain ideas related to land care in their mother tongues? We may find within Ghanaian languages, expressions, and cultural ideologies that explain aspects of the land or land care more richly than in English, especially in the papers of Kofi Agyekum and Robert Mate Wayo Opata.

Benjamin Abotchie Ntreh helpfully points to the Hebrew word, 'ereṣ, used for both "earth" and "land" and encompassing what is on, under, and above the earth. He also notes that the Ga people hold a similar view and for them, "land" therefore encompasses water, air, and earth. Although he does not mention the specific Ga terms, he suggests that Ga concepts could provide the avenue to develop a deeper understanding in conversations with people. Furthermore, the idea of land goes beyond the mere physical extant earth. Land also has to do with space and sacred spaces (Daniel Nii Aboagye-Aryeh), identity (Mark S. Aidoo), and a sense of belonging, both for humans and other-than-human creation.

The terms "environment" and "ecology" have distinctive meanings, although some people may use them interchangeably. The environment encompasses all that is around us. It has an anthropocentric emphasis. In other words, we usually use it to describe how humans relate to the created world.[1] Ecology, in its original usage, is literally the combination of eco (Greek οἶκος—house) and –logy (-λογία—the study of). So, it is the study of the interrelationship and interactions of organisms in their ecosystems and environment. The word "ecology" is now being used more broadly to describe different types of ecology, as in human ecology, political ecology, and so on.

Several of the authors provide an insightful discussion of the term "ecotheology," as we find in the papers by Mark S. Aidoo and Ebenezer Yaw Blasu. Blasu justifies his use of the term "theocology" in its focus on the study of the relationships between God as Creator and his creation. For a proper Ghanaian Christian view of land care, it is important not to assume that people understand these terms. Rather, we must provide adequate explanations before attempting to help them grow in their awareness of the problems related to land care.

---

1. See Moltmann, *Source of Life*, 120, 121.

2. Knowing the Problems Related to Land Care

It is clear from the papers that the authors have identified significant problems with respect to land care in both Ghana and Burkina Faso. Not only is human life endangered by human destruction, negligence, pollution, and interference with the land, but whole ecosystems have also broken down. Knowledge of the actual problems is a crucial part of constructing a Ghanaian Christian view and policy that informs effective land care. Below are a number of specific problems.

Desacralization of nature

In their discussion, both Ebenezer Yaw Blasu and Thomas Oduro make the reader aware that historically in Africa, there has been a desacralization of nature. This is reflected in the way the gospel has been negatively presented to impact the traditional rituals associated with sacred places and, by extension, the care of land and water. Nevertheless, there has been a continuity of traditional African thinking. For in Aboagye-Aryeh's discussion of the notion of wilderness, he reports that in Ghana, Christians still hold a view of the sacredness of place, particularly of mountains. However, it does not seem to carry over to land care, considering the way Ghanaians respond to the environment in general.

Separation between faith and land care (the environment)

The separation between faith and land care emerges clearly in Bonsu Osei-Owusu's paper. From his research in the Catholic church, he presents evidence that although the Catholic bishops in Ghana have, over a significant period of time, produced communiqués showing their awareness of the need to respond to environmental crises, few church members he sampled had ever heard of these policies. Furthermore, one keen Christian and active environmentalist regarded these "two aspects of her life" as "utterly unrelated to each other." Osei-Owusu is not alone with this finding.

When I asked Dr. Sherry Johnson, a Ghanaian Christian, veterinary surgeon, and lecturer at the University of Ghana for her view on land care, she responded: "I think we have separated Christianity from caring for the land. In my view, we think we can treat the land anyhow and then go to church on a Sunday and praise God all we want."[2]

---

2. Interview, Dr. Sherry Johnson, Oyarifa, 12 December 2018.

Evidently, some individuals and church denominations see land care and caring for creation as the sole responsibility of the Government or NGOs. In his research on Christian and traditional responses to mining in the Western, Eastern, and Ashanti Regions of Ghana, Christopher Affum-Nyarko found in 2014 and 2015, among Christians he interviewed from a range of church denominations, that church members had not heard any sermons addressing the problem of land care, the environment, or offering solutions to the problem of illegal mining.[3] Some church leaders even believed that "the church does not have any special authority or command to [be] involved in the fight against galamsey activities." They regarded Christian mission and galamsey mining as completely disparate activities. They maintained that the church should "discharge its spiritual responsibility of preparing souls for salvation while state agencies are resourced and empowered to fight galamsey mining."[4]

All the eighteen church ministers interviewed by Affum-Nyarko admitted that they scarcely ever spoke about environmental issues in their sermons or talks.[5] Not one of them had heard or themselves preached sermons responding to the issue of illegal mining and its related socioenvironmental problems.[6] Affum-Nyarko's stern warning is that:

> some leaders of the church see no interconnection between humans and the nonhuman creation of God as enshrined in Scripture. They do not see that the activities of illegal mining not only destroy creation, but can have serious detrimental effects on a person's neighbour whose drinking water he has spoiled. So they fail to see a connection with Jesus' command to 'love your neighbour as yourself.'[7]

Affum-Nyarko also discovered that proceeds from galamsey activity were going into building churches and other structures. He did however find some Christians who understood the link between faith and caring for creation. One Christian woman told him that "God is not separate from his creation," and thus, "When we destroy creation, we sin against the Creator." Some churches in the Eastern Region also occasionally organized gospel crusades to win miners (both legal and illegal) to Christ and teach them about land and water care. In general, Christian leaders had no discipleship programs that examined the church's response to illegal mining. In most of the Christian

---

3. Affum-Nyarko, "Theology, Human Need", 100, 103, 105.
4. Affum-Nyarko, "Theology, Human Need," 102.
5. See Affum-Nyarko, "Theology, Human Need," 100.
6. Affum-Nyarko, "Theology, Human Need," 104.
7. Affum-Nyarko, "Theology, Human Need," 100. Cf Matthew 22:39 ESV.

communities, Christians seemed quite indifferent to the problem. This leads us to the next problem of mining, particularly illegal mining.

Mining

Several articles in this volume discuss the problems of small-scale and illegal mining (galamsey). Adwoa Boadua Yirenkyi-Fianko raises certain issues concerning small-scale mining. John Appiah and James Kweku Agyen also discuss some of the problems resulting from galamsey mining, and explore possible solutions. Yaw Attah Edu-Bekoe details the hazards of galamsey mining, while Ini Dorcas Dah deplores the considerable destruction caused by small-scale mining in southwest Burkina Faso.

Affum-Nyarko reports evidence of the church's indirect complicity in illegal mining. One church elder was asked, "What can your local Assembly do to bring galamsey under control in your community?" He replied, "How can we preach against galamsey when the major financier to our church building project is a galamseyer?" He also indicated that the same person had contributed the greater part of the money used for acquiring musical instruments for the church.[8]

Quarrying and sand winning

None of the papers in this volume mentions these two activities as land care problems for Ghanaian Christians to consider their impact. For example, quarrying causes air pollution. Uncontrolled quarrying can destroy considerable areas of land and leave dangerous and unsightly water holes. The production of "clinker" from quarried limestone and clay used in the manufacture of cement produces large amounts of carbon dioxide (CO_2).[9] This, in turn, influences climate change and impacts global warming, which affects all of creation.

Deforestation

A few authors touch upon the issue of deforestation in Ghana (Osei-Owusu, Appiah, and Agyen) or allude to it in their discussion on illegal mining. Yet, deforestation is a major problem in Ghana. A number of causes of deforestation are associated with land care: extension of agricultural land, charcoal

8. Affum-Nyarko, Email communication, 31 May 2017.
9. Rodgers, "Massive CO2 Emitter."

burning, bush burning, mining, disposal of waste (particularly toxic waste), and the extensive boom in the construction of buildings.

Pollution: plastics; open defecation, general waste; noise pollution; air pollution

In his discussion of pollution, Yaw Attah Edu-Bekoe raises a range of problems related to landfill sites, particularly from techno-waste or e-waste, and their highly toxic impact on everything in creation. He also discusses air pollution and touches on the problems of plastics. Plastic waste is a major pollutant in Ghana, choking both land and water bodies. Human attitudes, including those of Christians, toward rubbish disposal in general, is significantly problematic and this has profound implications for land care.

Bush burning

Haruna Mogtari briefly mentions the problems Fulani herdsmen have with people burning bushes and the destruction of trees, grassland, and food sources for their cattle. Although a predominantly rural problem, this is significant for land care. Bush burning not only exposes the soil to erosion, but it also endangers plant, animal, and insect species, as well as human life. One of my saddest experiences in Ghana was trying to condole with a Ghanaian woman whose son, a Christian pastor, died as a result of a bush fire set to hunt animals more easily. Things went badly wrong when, with a sudden change in wind direction, he was engulfed in flames and was badly burnt. He died later in the hospital.

Land use conflict

Mogtari also emphasizes the significant problem in Ghana of land use conflict, particularly between cattle herders and farmers. Invariably, this has an impact on land care. Ini Dorcas Dah shows how gold mining results in land use conflict in Burkina Faso. The spread of buildings is a major cause of tension over land use and ultimately land care. In some rural areas, the numbers of buildings are rapidly increasing and farmers are increasingly concerned about being "pushed off" their farmland, with little or no compensation.

Uncontrolled building works in both urban and rural areas

The rate at which buildings are constructed and small estates developed has rapidly increased in Ghana since the 1980s. Although building and land use regulations are in place, enforcement is significantly difficult. In some areas, regulations related to land use are nonexistent. Throughout the country, designating green spaces and preserving small clumps of forest on the outskirts, within, and through cities does not appear to be a high priority. Land developers bulldoze everything in sight rather than consider preserving trees that would not interfere with buildings. Some developers also bulldoze the topsoil off the land in order to construct buildings, then the new house owners have to buy black soil to construct a garden. This contributes to the problem of stealing soil. As Gideon Jampana, a Christian laborer who has worked in the building industry, told me, "When we were doing construction work, a lot of people also steal soil. They will go with a car into the bush somewhere and they will dig the land and make big holes there and leave it. They take the soil illegally and go and sell it."[10] With land being sold now in Ghana at increasingly exorbitant prices, careful planning of land use in both city and rural areas cannot be neglected.

Climate Change

Although none of the authors in this volume directly referred to climate change, some like Edu-Bekoe mention the depletion of the ozone layer. As he rightly points out, this contributes to global warming. However, climate change goes beyond just the depletion of the ozone layer. Many of the above activities contribute to climatic change, which in turn will further affect land care.

Herbicides, Weedicides, Insecticides, Fertilizers

Not one author mentioned the problems that can arise from using herbicides, weedicides, insecticides, and fertilizers in land care. Although all of these products may have benefits in land use and food production, they also have negative long-term effects on land, water and human health.

---

10. Interview on a Christian view of land care, Gideon Jampana, Oyarifa, Accra, 12 December 2018.

3. Understanding the Wide-ranging Impact

It is imperative that a Ghanaian Christian view of land care include a clear understanding of the wide-ranging impact of the above problems, not just on humans, but also on all of creation. Understanding is a gift of God that requires our knowledge of the problem and divine help to frame better responses. I have grouped the range of impact under themes.

Physical

Many of the listed problems have physical consequences. This is particularly evidenced in Edu-Bekoe's paper and in those that examine small-scale mining. Ini Dorcas Dah's clear picture of the shocking land degradation occurring in the name of short-term profits in Southwest Burkina Faso is an example.

There are physical consequences for environmental health in general. All of creation is impacted. It affects the health of entire ecosystems––humans, animals, birds, fish, and all life. Plastic waste has become the scourge of land and water. Not only are animals and fish ingesting plastic matter now, it has also been found in humans.[11]

Dr. Sherry Johnson is currently doing research on dogs in the Kyebi area of Ghana's Eastern Region. She is collecting canine blood samples to see the levels of mercury, lead, and cadmium in them. These three heavy metals used in mining are extremely injurious to human health. Used in the washing of gold ore, mercury is extremely toxic to both humans and animals. It is both ingested and inhaled into the lungs. Dr. Johnson explained that dogs "are used as sentinels because they respond to environmental pollutants, including heavy metals, just the way we do." Also, because dogs "have a shorter life span [than humans], they can develop chronic diseases like cancer much quicker than we do." Testing dogs gives "a very good estimation of what human beings are being exposed to but probably are not showing any signs of at the moment." She indicated that the levels of mercury she is finding in dogs "right now is about 20 to 30 percent higher than what it should be and it is scary."[12]

During her research in Kyebi, Dr. Johnson found that people had been paid to move out of their homes and the mining was being done right under those homes. In the short term, mining is highly profitable, but, according to her, most people in the community are "probably unaware of the injury

---

11. Harvey and Watts, "Microplastics Found in Human Stools."
12. Interview, Dr. Sherry Johnson, Oyarifa, 12 December 2018.

these heavy metals cause to their health." Dr. Johnson pointed out that the renal center at Korle Bu Hospital is overwhelmed with sick people with kidney problems. Cancer cases are also on the rise.

Social

The problems in the previous section impact society, each in a different way. Ini Dorcas Dah brings out the insecurity in neighboring Burkina Faso that has resulted from gold mining in communities, with increasing incidents of armed robbery. This is occurring in Ghana as well. Dah also reveals the very negative impact of gold mining on the education of children and teenagers, with prostitution on the increase. From both personal observation and his interactions with traditional elders, Professor Robert Addo-Fening noted the following trend in his home area of Osino, in Ghana's Eastern Region, where teenagers flock to mining:

> The involvement of children in mining not only means that they do not go to school, but has led to increasing immorality and disrespect toward parents and the elderly. Children and teenagers with access to money from gold do not necessarily use this for their education. It is creating a class of children and teenagers who have the ability to go to market and buy food for themselves rather than eat with their parents. The level of insolence is shown where teenagers through what they earn from gold and sakawaa can afford a car. So why should they respect an elderly person who doesn't even own a bicycle?[13]

The loss of farmland to buildings can create enormous social problems if families are not compensated adequately either by good alternative farmland or by (re-)training for new areas of employment. In a Ghanaian Christian view of land care, an understanding of the social consequences is imperative for appropriate response. This means that Ghanaian Christians need to do research in order to study and understand these consequences.

Economic/material

In a Ghanaian Christian view of land care, understanding the economic or material impact of the problems mentioned in the second section reveals that the economic use of land can both create wealth and impoverish

---

13. Addo-Fening, Personal Communication, 28 May 2017.

people. Ini Dorcas Dah in her article examines this economic impact in Burkina Faso.

Religious/spiritual/aesthetic

Bonsu Osei-Owusu brings out the failure of church leaders to communicate effectively to those at the grassroots level because their message has no impact. Most of the other papers in the volume reflect on various biblical aspects related to land care. It is therefore critical to understand that failure to care for the land has not only aesthetic and spiritual consequences, but a religious impact as well. A number of the authors refer to destructive practices as "sin." Failure to protect and manage the life resources of creation in a wise way that sustains all creation is tantamount to sin against God. This has a huge impact on our spiritual life.

Political

The final area of impact is politics. Some of the authors here refer to government activities in land care. They also point to the inadequacies or lack of action. Where destructive practices related to land care generate wealth, they attract people in positions of power. Even though they should know better, they either turn a blind eye to what is happening, or participate for personal profit. Others, however, who act to prevent destructive practices, have experienced considerable personal and political risk.

4. Developing a Strategic Ghanaian Christian Response to Land Care

The fourth aspect of having a Ghanaian Christian view of land care is the development of a strategic Ghanaian Christian response to the issues and problems that Christians have understood. The papers in this volume show that some Ghanaian Christians are making a clear effort to address issues of land care by bringing them to the attention of both individuals and Christian institutions. This would not only elicit a Christian response, but also identify land care as a highly significant training activity at all levels of church and society.

In conjunction with identifying land care problems and understanding the themes as illustrated by the wide-ranging impact noted above, it is critical that a Ghanaian Christian view of land care include the holistic knowledge and understanding of biblical themes. This helps us develop a spiritual perspective which, when combined with knowledge of other responses, provides a basis for all practical response. A personal theology that separates the physical from the spiritual leads to a distorted spirituality.

Theological, Religious, Spiritual

From the authors' discussions, a number of distinctly interlinked biblical themes emerge with implications for a Christian view of land care, namely, God/Supreme Being, creation, community, sinfulness of environmental destruction, land/water/air, salvation/redemption/reconciliation, and renewal of the earth. For each of these themes, a sound theological understanding is required.

a. God/Supreme Being

The papers by Blasu, Appiah and Agyen, Edu-Bekoe, and Osei-Owusu stress the importance of a theocentric approach to land care. God as the Creator is the one who is over all creation. This understanding brings into better perspective the place of humans within creation as part of the created world, yet with a divine responsibility. Blasu proposes an African theocology that proposes a study of ecology that begins with God, just as God's mission begins with God. He indicates that it is "a holistic study of our natural home, the Earth, and our moral and missional responsibility to it."

b. Creation

Other authors focus on different aspects of the creation stories in Genesis 1 and 2. For example, Ntreh and Twum-Baah look at the misinterpretation of the terms "dominion" and "subdue" in Genesis 1:28 (see also the articles by Edu-Bekoe and Dah) and the fact that these terms are not meant for exploitation, but rather blessing. Aidoo underscores creation as a life concern (Genesis 2:4b–17) and connects the passage to the Akan view of human identity linked to the land. He emphasizes human stewardship and the role of a servant of the ground. Edu-Bekoe also supports a "cultural mandate of stewardship." Blasu gives prominence to humanity created in the image of God and the implications for land care. He also points to the human kinship with nature that emerges in both the creation stories and other parts of Scripture, and links this with African worldviews. This human/nature relationship has implications for understanding community in creation.

c. Community

A theology of community and its relationship with land is not directly addressed in any of the papers, yet this has a direct bearing on issues of land care. In Genesis 1 and 2, we see the first communities of creation as well as the relationships within a group and between groups. For example, the earth and sky; the sun, moon and stars; land and water (the hydrological system); land and vegetation; animals and humans (living creatures—*nephesh hayah*) and the relationships that exist among all the creatures on the earth.

In Genesis 3, we see the disruptions of all the relationships caused by human sin. Community is thus disrupted. Ntreh discusses open defecation and God's mind in the instructions he gave to Israel (Deut 23). He prescribed a preventative measure that did not only have aesthetic value, but was also hygienic and counteracted the spread of disease and any subsequent disruption in community life through occurrences such as cholera outbreaks.

Aidoo, for his part, stresses that "stewardship goes beyond individual gifts to encompass the whole of creation." He brings out the importance of community in the way in which people care for creation and in the "values of life present in [the] African culture of communion." These are central biblical values for a Ghanaian Christian view of land care.

Other authors, like Appiah and Agyen, while not directly using the word "community," allude to it in discussing stewardship. Edu-Bekoe supports the position that ecology is about relationships and implies that this is about community. Therefore, when people break God's command and destroy the earth, it disrupts community. Mogtari discusses Scripture's mandate for people to care for the foreigner, and points out the consequences for land care with regard to herder-farmer relationships.

d. Sinfulness of environmental destruction

Osei-Owusu found that the Christians in his study came to a consensus that degrading the environment was tantamount to sin. Using Matthew 6:3, Blasu argues that "failure to let God's concern for his environment be central in reducing deforestation could be classified as evangelical sin." This aligns with what is portrayed in Genesis 3 where all the communal relationships that God created become disrupted. Degradation of the environment is a profound disruption of the relationship between humans and land, water, and air. Actions that degrade both human and other-than-human creation can be classified as sin. This has great implications where activities such as mining take place and the destroyed environment is not renewed.

e. Land / water / air

A number of the authors discuss Scripture's perspective on land, water, and air. Care for the land means treating the occupants of the land well: the living creatures, as well as the trees, the land itself, water, and the air. There is potential for further development of a theology of land from a Ghanaian Christian perspective. Oduro shows that traditional religious approaches to the land/earth have theological significance and find their continuity in a Christian response to the environment. Thus, he expounds a theology of mountains and a theology of water bodies.

Daniel Nii-Aboagye Aryeh provides a helpful theological study of ἔρημος *hérēmos* ("wilderness, desolate place") particularly from the Gospel of Luke. He places it in the context of the Old Testament and the intertestamental period to focus on sacred space in Ghana and sacred spaces in African Christianity. He calls us to identify mountains as sacred spaces. Mogtari highlights passages from Scripture related to land and water conflicts. This is pertinent for a holistic response to the herder-farmer issues in Ghana.

The understanding of the sacredness of land and space could be further expanded beyond just mountains and water. For if the earth is the Lord's, then there is a sacred component to all the earth and all that is in, on, under, and above it. This has implications for Ghanaian Christians caring for land through a deliberate development of aesthetically beautiful "dedicated spaces" in cities and communities where people can go to meditate and pray, as well as just to enjoy the beauty.

f. Salvation / Redemption / reconciliation

One area often lacking in a Ghanaian Christian view of land care is an adequate theology of salvation, redemption, and reconciliation with respect to land care and the environment. Edu-Bekoe brings out the mandate given to Jesus' disciples to "preach the good news to all creation" (Mark 16:15). He explains that "while we are doing our part rescuing the perishing, we should be able to salvage or redeem the degraded ecology." Blasu refers to the passage in Colossians 1:19-20—especially verse 20—where Paul emphasizes the reconciliation of all creation in heaven and on earth through Christ's blood shed on the cross. This has profound implications for a Ghanaian

Christian view of land care. It means that the land and everything related to it comes under the purview of Jesus' reconciling power. Therefore, as Christians we cannot ignore our responsibility to act as reconcilers in the universe. We have a role as guardians and coworkers in conserving that for which Christ died. Blasu also rightly points out that even within creation, there are "biophysical processes" that serve to provide "divine healing." Our role is to understand and reinforce this.

Ntreh notes the link in "The Lord's Prayer" of heaven and earth with God's will, pointing to God's interest in all creation. The purpose of the coming of God's kingdom is to renew all creation. Oduro refers to the insightful ways that Afua Kuma very clearly links Jesus with all creation in her prayers and praises.

g. Renewal of the earth

None of the authors discuss the meaning of the "new earth." I have personally heard Christians in Ghana use this verse to justify their not caring for the environment or land. They claim that since God will one day destroy the earth and replace it with a completely new earth, why bother?

It is critical to understand that the Greek word used for "new" is καινός *kainós* and not νέος *néos*. καινός implies that something is better because it is different.[14] So, it has the meaning of renewal rather than replacement with something that has never existed before. In contrast, νέος *néos* means something that has recently come into existence or is created recently in time. In Revelation 21:1, the word is *kainós*, meaning a renewed earth and heavens.

h. Role of the Church in land care

The above themes all point to the church's role in land care. Ini Dorcas Dah underscores this when she calls for the church to hold a clear position related to the abuse of the land she observed. Osei-Owusu provides a clear discussion about the church's role and offers practical suggestions on how the church can act in a ministry of land care. The church cannot ignore land care because climate change will have significant consequences on the land, which will in turn impact church ministries.[15] This would require developing a program of practical education at all levels of church ministry, to all

---

14. Zodhiates, *Complete Word Study Dictionary*.
15. Edgar and Xu, "Climate and Church."

age groups, and in all formats of church activities. It also means developing practical activities, such as tree planting and waste management programs (some churches have plastic waste littered all over their premises).

Social

A Ghanaian Christian view of land care involves a response to social problems related to the misuse of land. A number of the authors here address issues such as community-based approaches (Yirenkyi-Fianko), tree planting, youth-driven focus, attitudinal changes in youth (Edu Bekoe), and responding to immorality (Dah). Where people are displaced because of land care issues, the Christian's role is to both confront the issues that caused the displacement and provide care for the displaced persons. That is part of loving our neighbor.

Political

A Ghanaian Christian view of land care includes an awareness of the activities of government and secular organizations related to land care. Yirenkyi-Fianko names some governmental agencies and other organizations involved in water management and points to some of the studies that have been undertaken. Edu-Bekoe notes the strong stances some governments have adopted in Africa to deal with environmental degradation. He calls for the endorsement, ratification, and enforcement of international communiqués, and highlights the difficulty that the Ghanaian government faces in enforcing sanctions against those who cause environmental degradation of the land.

As Osei-Owusu points out, the church cannot continue to remain silent. She needs to use her prophetic voice of advocacy with the government and society to support the implementation of helpful programs, critique observable structural sin, and show leadership through creative conservation of the land.

Economic / material

In the history of Christianity, Christians have been noted for their giving and practical action related to land care issues.[16] However, sometimes Christians, due to greed or misguided economic actions, can contribute

16. Robert, "Historical Trends in Missions," 123–28.

to exacerbating land care problems. Only a few churches in Ghana participate in plastic-waste recycling programs.[17] Edu-Bekoe calls for waste management companies to help in the process by providing separate receptacles for different types of waste. As indicated above, plastic waste has become the scourge of land and sea. A ban on the use of plastics would have an economic impact on plastic producers. Even so, the economic impact of the consequences of plastic waste with respect to land care, human, and environmental health is already significant. Ghanaian Christian scientists have a responsibility to lead the way in the search for alternatives. But, for all of us Ghanaian Christians in our view of land care, do we stop to think about how we can use our God-given resources and finances to bring healing to the land?

Physical

So far, I have touched upon physical things that Ghanaians could do about land care. A Ghanaian Christian view of land care means that Christians must be physically involved in practical actions that contribute to land care. This would require a change in both personal and public attitudes. As Dah notes in her paper, a Ghanaian Christian view of land care requires involvement in education in both the church and secular institutions. As Aidoo points out, it calls for an inclusive education where all theological subjects are ecotheological in approach, and are also related to relevant secular subjects.

Conclusion

When I first arrived in Ghana in 1981, I drove the length of Ghana through pristine rainforests in the south and a tree-filled savannah in the north. I was struck by the cleanliness of Accra and Kumasi, with people sweeping the little areas in front of their homes and shops every day. The beaches were pristine and in the town of Buipe, I could see the clean shores from the banks of the Black Volta River. Sadly, by 2019, all this has changed. The forests have mostly disappeared along the route. The trees of the savannah are by far fewer and the pollution in some places makes you weep, especially when you see how it is destroying beaches and water bodies.

---

17. One notable example of a church with a commendable plastic-waste recycling program is the Legon Interdenominational Church, Legon, on the University of Ghana campus.

Approximately 2,000 years ago, the apostle John, in Revelation 5:13, heard "every creature in heaven and on earth and under the earth and in the sea, and all that is in them, singing, 'To the one seated on the throne and to the Lamb, be blessing, and honor and glory and might for ever and ever.'" If we fail to care for the land and some of these creatures become extinct, how can they sing praises to God, to Christ? William Barclay writes, "Here is the truth that heaven and earth and all that is within them is designed for the praise of Jesus Christ; and it is our privilege to lend our voices and our lives to this vast chorus of praise, for that chorus is necessarily incomplete so long as there is one voice missing from it."[18] A Ghanaian Christian view of land care necessarily requires concerted practical action. Therefore, every Ghanaian Christian must be personally committed to the renewal of the land, water, and air as part of a covenant before God.

Bibliography

Barclay, William. *The Revelation of John*. Volume 1. Philadelphia: Westminster John Knox, 1976.

Edgar, Tylor, and Lee Xu. "Climate and Church: How Global Climate Change Will Impact Core Church Ministries." https://www.interfaithpowerandlight.org/wp-content/uploads/2009/11/ClimateWhitePaper_finalREV.pdf.

Harvey, Fiona, and Jonathan Watts. "Microplastics Found in Human Stools for the First Time." https://www.theguardian.com/environment/2018/oct/22/microplastics-found-in-human-stools-for-the-first-time.

Moltmann, Jürgen. *The Source of Life: The Holy Spirit and the Theology of Life*. Minneapolis: Fortress, 1997.

Nyarko, Christopher Affum. "Theology, Human Need and the Environment: An Evaluation of Christian and Traditional Responses to Illegal Mining in Ghana." MTh diss., Akrofi-Christaller Institute of Theology, Mission and Culture, Akropong-Akuapem, Ghana, 2016.

Robert, Dana L. "Historical Trends in Missions and Earth Care." *International Bulletin of Missionary Research* 35.3 (July 2011) 123–28.

Rodgers, Lucy. "The Massive CO2 Emitter You May Not Know About." https://www.bbc.com/news/science-environment-46455844.

Zodhiates, Spiros. *The Complete Word Study Dictionary: New Testament*, Kindle. Chattanooga, TN: AMG, 2000.

---

18. Barclay, *Revelation of John*, 173.

www.ingramcontent.com/pod-product-compliance
Lightning Source LLC
Chambersburg PA
CBHW050844230426
43667CB00012B/2141